NANOTECHNOLOGY

NANOTECHNOLOGY

Manoj Bhatia

ANMOL PUBLICATIONS PVT. LTD.
NEW DELHI-110 002 (INDIA)

ANMOL PUBLICATIONS PVT. LTD.
Regd. Office: 4360/4, Ansari Road, Daryaganj,
New Delhi-110002 (India)
Tel.: 23278000, 23261597, 23286875, 23255577
Fax: 91-11-23280289
Email: anmolpub@gmail.com
Visit us at: www.anmolpublications.com

Branch Office: No. 1015, Ist Main Road, BSK IIIrd Stage
IIIrd Phase, IIIrd Block, Bangalore-560 085 (India)
Tel.: 080-41723429 • Fax: 080-26723604
Email: anmolpublicationsbangalore@gmail.com

Nanotechnology

First Edition, 2010

ISBN 978-81-261-4508-9

PRINTED IN INDIA

Printed at Mehra Offset Press, Delhi.

Contents

Preface

The book is self-contained and unified in presentation. It may be used as a textbook by graduate students as well as ambitious undergraduates in engineering, biological, and physical sciences who have some familiarity with quantum and statistical mechanics. It is also suitable for experts in the related fields who require an overview of the fundamental topics in nanotechnology. The explanations in the book are detailed enough to capture the interest of the curious reader, and complete enough to provide the necessary background material needed to go further into the subject and explore the research literature. Nanotechnology is the design, characterization, production and application of structures, devices and systems by controlling shape and size at the nanoscale.

This book provides an excellent introduction to the theories and experimental methods of molecular based study of condensed matter in small systems which are the principles of nanotechnology. For new comers to the field it is a marvelous and thorough educational tool. For those already familiar with some of the concepts, it makes more concrete their knowledge for possible earlier stages entering into research and development in nanoscience and nanotechnology.

Author

Chapter 1

Introduction

NANO SCIENCE AND NANOTECHNOLOGY

DEFINITION

Nanotechnology is the engineering of functional systems at the molecular scale. This covers both current work and concepts that are more advanced. In its original sense, 'nanotechnology' refers to the projected ability to construct items *from the bottom up*, using techniques and tools being developed today to make complete, high performance products.

With 15,342 atoms, this parallel-shaft speed reducer gear is one of the largest nanomechanical devices ever modeled in atomic detail.

MEANING OF NANOTECHNOLOGY

When K. Eric Drexler popularized the word 'nanotechnology' in the 1980's, he was talking about building machines on the scale of molecules, a few nanometers wide—motors, robot arms, and even whole computers, far smaller than a cell. Drexler spent the next ten years describing and analyzing these incredible devices, and responding to accusations of science fiction.

Meanwhile, mundane technology was developing the ability to build simple structures on a molecular scale. As nanotechnology became an accepted concept, the meaning of the word shifted to encompass the simpler kinds of nanometer-scale technology. The National Nanotechnology Initiative was created to fund this kind of nanotech: their definition includes anything smaller than 100 nanometers with novel properties.

Much of the work being done today that carries the name 'nanotechnology' is not nanotechnology in the original meaning of the word.

Nanotechnology, in its traditional sense, means building things from the bottom up, with atomic precision. This theoretical capability was envisioned as early as 1959 by the renowned physicist Richard Feynman.

I want to build a billion tiny factories, models of each other, which are manufacturing simultaneously. The principles of physics, as far as I can see, do not speak against the possibility of maneuvering things atom by atom. It is not an attempt to violate any laws; it is something, in principle, that can be done; but in practice, it has not been done because we are too big.Richard Feynman, Nobel Prize winner in physics Based on Feynman's vision of miniature factories using nanomachines to build complex products, advanced nanotechnology (sometimes referred to as molecular manufacturing) will make use of positionally-controlled mechanochemistry guided by molecular machine systems.

Formulating a roadmap for development of this kind of nanotechnology is now an objective of a broadly based technology roadmap project led by Battelle (the manager of several National Labouratories) and the Foresight Nanotech Institute.

Shortly after this envisioned molecular machinery is

created, it will result in a manufacturing revolution, probably causing severe disruption. It also has serious economic, social, environmental, and military implications.

GENERATIONS

Mihail Roco of the National Nanotechnology Initiative has described *four generations* of nanotechnology development. The current era, as Roco depicts it, is that of passive nanostructures, materials designed to perform one task. The second phase, which we are just entering, introduces active nanostructures for multitasking; for example, actuators, drug delivery devices, and sensors.

The third generation is expected to begin emerging around 2010 and will feature nanosystems with thousands of interacting components. A few years after that, the first integrated nanosystems, functioning (according to Roco) much like a mammalian cell with hierarchical systems within systems, are expected to be developed.

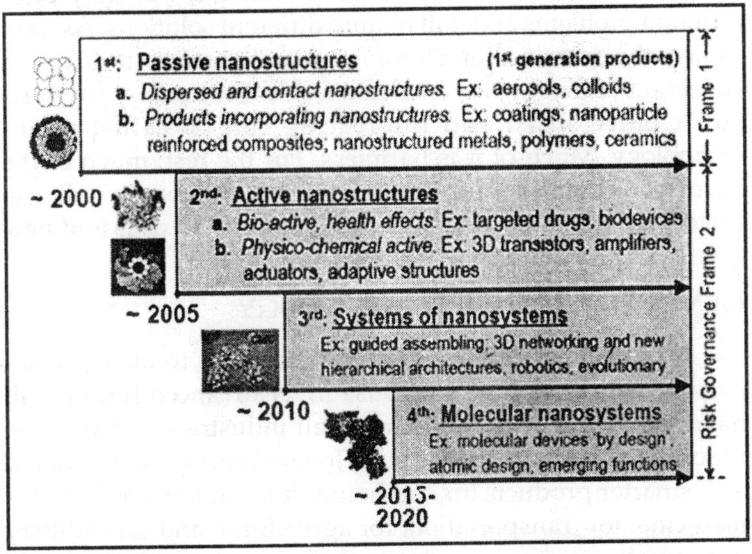

Some experts may still insist that nanotechnology can refer to measurement or visualization at the scale of 1-100 nanometers, but a consensus seems to be forming around the idea that *control and restructuring of matter* at the nanoscale is a necessary element. CRN's definition is a bit more precise than that, but as work progresses through the four generations of nanotechnology leading up to molecular nanosystems, which will include molecular manufacturing, we think it will become increasingly obvious that "engineering of functional systems at the molecular scale" is what nanotech is really all about.

CONFLICTING DEFINITIONS

Unfortunately, conflicting definitions of nanotechnology and blurry distinctions between significantly different fields have complicated the effort to understand the differences and develop sensible, effective policy.

The risks of today's nanoscale technologies (nanoparticle toxicity, etc.) cannot be treated the same as the risks of longer-term molecular manufacturing (economic disruption, unstable arms race, etc.).

It is a mistake to put them together in one basket for policy consideration each is important to address, but they offer different problems and will require different solutions. As used today, the term nanotechnology usually refers to a broad collection of mostly disconnected fields. Essentially, anything sufficiently small and interesting can be called nanotechnology. Much of it is harmless. For the rest, much of the harm is of familiar and limited quality. But as we will see, molecular manufacturing will bring unfamiliar risks and new classes of problems.

GENERAL-PURPOSE TECHNOLOGY

Nanotechnology is sometimes referred to as a *general-purpose technology*. That's because in its advanced form it will have significant impact on almost all industries and all areas of society. It will offer better built, longer lasting, cleaner, safer, and smarter products for the home, for communications, for medicine, for transportation, for agriculture, and for industry in general.

Imagine a medical device that travels through the human body to seek out and destroy small clusters of cancerous cells before they can spread. Or a box no larger than a sugar cube that contains the entire contents of the Library of Congress. Or materials much lighter than steel that possess ten times as much strength. — U.S. National Science Foundation.

DUAL-USE TECHNOLOGY

Like electricity or computers before it, nanotech will offer greatly improved efficiency in almost every facet of life. But as a general-purpose technology, it will be *dual-use,* meaning it will have many commercial uses and it also will have many military uses—making far more powerful weapons and tools of surveillance. Thus it represents not only wonderful benefits for humanity, but also grave risks.

A key understanding of nanotechnology is that it offers not just better products, but a vastly improved manu-facturing *process*. A computer can make copies of data files—essentially as many copies as you want at little or no cost. It may be only a matter of time until the building of products becomes as cheap as the copying of files. That's the real meaning of nanotechnology, and why it is sometimes seen as "the next industrial revolution."

My own judgment is that the nanotechnology revolution has the potential to change America on a scale equal to, if not greater than, the computer revolution. — U.S. Senator Ron Wyden (D-Ore.)

The power of nanotechnology can be encapsulated in an apparently simple device called a *personal nanofactory* that may sit on your countertop or desktop. Packed with miniature chemical processors, computing, and robotics, it will produce a wide-range of items quickly, cleanly, and inexpensively, building products directly from blueprints.

EXPONENTIAL PROLIFERATION

Nanotechnology not only will allow making many high-quality products at very low cost, but it will allow making new nanofactories at the same low cost and at the same rapid speed. This unique (outside of biology, that is) ability to reproduce its own means of production is why nanotech is said to be an *exponential* technology. It represents a manufacturing system that will be able to make more manufacturing systems—factories that can build factories—rapidly, cheaply, and cleanly. The means of production will be able to reproduce exponentially, so in just a few weeks a few nanofactories conceivably could become *billions*. It is a revolutionary, transformative, powerful, and potentially very dangerous—or beneficial—technology.

How soon will all this come about? Conservative estimates usually say 20 to 30 years from now, or even much later than that. However, CRN is concerned that it may occur sooner, quite possibly within the next decade. This is because of the rapid progress being made in enabling technologies, such as optics, nanolithography, mechano-chemistry and 3D prototyping. If it does arrive that soon, we may not be adequately prepared, and the consequences could be severe.

We believe it's not too early to begin asking some tough questions and facing the issues:

- Who will own the technology?
- Will it be heavily restricted, or widely available?
- What will it do to the gap between rich and poor?
- How can dangerous weapons be controlled, and perilous arms races be prevented?

Many of these questions were first raised over a decade ago, and have not yet been answered. If the questions are not answered with deliberation, answers will evolve independently and will take us by surprise; the surprise is likely to be unpleasant.

It is difficult to say for sure how soon this technology will mature, partly because it's possible (especially in countries that do not have open societies) that clandestine military or industrial development programmes have been going on for years without our knowledge.

We cannot say with certainty that full-scale nanotechnology will not be developed with the next ten years, or even five years. It may take longer than that, but prudence—and possibly our survival—demands that we prepare now for the earliest plausible development scenario.

PRODUCTS FROM ATOMS

The properties of those products depend on how those atoms are arranged. If we rearrange the atoms in coal we can make diamond. If we rearrange the atoms in sand (and add a few other trace elements) we can make computer chips. If we rearrange the atoms in dirt, water and air we can make potatoes.

Todays manufacturing methods are very crude at the molecular level. Casting, grinding, milling and even lithography move atoms in great thundering statistical herds. It's like trying to make things out of LEGO blocks with boxing gloves on your hands. Yes, you can push the LEGO blocks into great heaps and pile them up, but you can't really snap them together the way you'd like.

In the future, nanotechnology will let us take off the boxing gloves. We'll be able to snap together the fundamental building blocks of nature easily, inexpensively and in most of the ways permitted by the laws of physics. This will be essential if we are to continue the revolution in computer hardware beyond about the next decade, and will also let us fabricate an entire new generation of products that are cleaner, stronger, lighter, and more precise.

It's worth pointing out that the word "nanotechnology" has become very popular and is used to describe many types of research where the characteristic dimensions are less than about 1,000 nanometers. For example, continued improvements in lithography have resulted in line widths that are less than one micron: this work is often called "nanotechnology." Sub-micron lithography is clearly very valuable (ask anyone who uses a computer!) but it is equally clear that conventional lithography will not let us build semiconductor devices in which individual dopant atoms are located at specific lattice sites. Many of the exponentially improving trends in computer hardware capability have remained steady for the last 50 years. There is fairly widespread belief that these trends are likely to continue for at least another several years, but then conventional lithography starts to reach its limits.

If we are to continue these trends we will have to develop

a new manufacturing technology which will let us inexpensively build computer systems with mole quantities of logic elements that are molecular in both size and precision and are interconnected in complex and highly idiosyncratic patterns. Nanotechnology will let us do this.

When it's unclear from the context whether we're using the specific definition of "nanotechnology" (given here) or the broader and more inclusive definition (often used in the literature), we'll use the terms "molecular nanotechnology" or "molecular manufacturing."

Whatever we call it, it should let us:

- Get essentially every atom in the right place.
- Make almost any structure consistent with the laws of physics that we can specify in molecular detail.
- Have manufacturing costs not greatly exceeding the cost of the required raw materials and energy.

POSITIONAL ASSEMBLY

Introduction

Manufactured products are made from atoms. The properties of those products depend on how those atoms are arranged. Viewed from the molecular level today's macroscopic manufacturing methods are crude and imprecise. Casting, milling, welding and all the other traditional manufacturing methods spray atoms about in great statistical herds.

Even lithography (which already lets us put millions of transistors on a chip no bigger than your fingernail) is fundamentally statistical and random. Exactly how many dopant atoms are in a single transistor and exactly where each individual dopant atom is located is neither specified nor known: if we have roughly the right number in roughly the right place, we can make a working transistor. For today, that is good enough.

The exception is chemistry. Large high purity crystals can have almost every atom in the right place. So, too, can many long polymers. The structures of proteins with hundreds and even thousands of amino acids can be specified down to the last atom.

Most dramatically (and fortunately for us!) DNA strands with many tens of millions of bases can be copied with almost perfect accuracy. And it seems that almost any small molecule (with perhaps several dozens of atoms) can be synthesized, if only we have the skill and patience. Yet the laws of physics and chemistry in principle permit arranging and rearranging the elements in so many combinations and permutations that all of our manufacturing skills and all of our chemical skills barely suffice to scratch the surface of what is possible.

Utility of Diamond

Almost any manufactured product could be improved, often by several orders of magnitude, if we could precisely control its structure at the molecular level. We often want our products to be light and strong. Diamond is light and strong: the strength-to-weight ratio of diamond is over 50 times that of steel.

Yet we do not today have diamond spars in airplanes nor diamond hulls for rockets. Today we can't economically make diamond. Even if we could, simple diamond crystals can shatter. We'd have to modify the structure to make it tough and shatter proof: perhaps diamond fibers. While easily done in principle, we can't do this in practice today.

Great strength and light weight are not the exclusive province of diamond: graphite can be stronger. And if we consider the many ways in which carbon atoms can be arranged and rearranged, then it's obvious that there are a host of other possibilities. Yet all share a common problem: we can't yet economically make them in the exact shapes that we want.

Great strength is only one property that we prize highly: when we make computers we are more concerned by electrical properties. Here, too, diamond excels. Today's computers are made of semiconductors, and the semiconductor of choice is silicon.

This is not because silicon is the ideal semiconductor from which to make computers, but because we know how to make devices from it. The computer industry has strong opinions about what makes a good logic device and what makes a good computer and diamond will let us make better computers than silicon . Diamond has a wider bandgap, hence electrical devices will work at higher temperatures. It has greater thermal conductivity, so devices can be more easily cooled.

It has a greater breakdown field, hence devices can be smaller. It has higher electron and hole mobility which, when combined with higher electric fields, will result in higher speed. But again, we see no diamond computers, just as we see no diamond airplanes: we can't economically manufacture them yet. Large pure crystals of silicon can be made relatively easily, but large pure crystals of diamond are scarce. We can etch the silicon surface and add dopants with a precision measured in tenths of microns, while the corresponding steps for diamond are more difficult. Not more difficult in principle: just more difficult today.

Long Range Complex Order

Making computers highlights another problem. It's not enough to make a pure crystal, it must also have an extremely precise and complex pattern of impurities. The exact location of the dopant atoms in the semiconductor lattice controls how devices function and where signals can propagate. Local order is crucial to make each device work, but long range complex order is crucial to make the computer as a whole work. While we can make some things today that are highly precise and have simple long range order (e.g., crystals), it is the requirement for complex long range order that prevents us from making computers of the kind we'd like to make. While it's plausible we could make high density memory from crystals and perhaps some types of cellular automatons, we couldn't make anything that resembled the computers on the market today. Today's high speed semiconductor-based digital computers (like the 80486 or the Pentium) have millions of logic elements wired together in complex and highly idiosyncratic patterns. This is well beyond the capabilities of crystal growth or bio-polymer synthesis. It will require a fundamentally new manufacturing technology: molecular manufacturing.

A Gap

Today, there is a gap in our synthetic abilities: we can make complex mechanical machinery and electronic devices (including computers, which have millions of transistors), but we can't make such devices with the precision with which the chemist can synthesize a crystal, a bio-polymer, or a relatively small molecule. With chemistry we can make precise molecular structures and compounds, but we haven't been able to scale up that success to molecular computers (and other macroscopic products as precise as molecules).

Molecular manufacturing will, by definition, let us economically manufacture almost any specified structure that is consistent with the laws of chemistry and physics. To simplify the problem somewhat we can narrow our focus to structures that resemble diamond in a broad sense: the diamondoid structures as defined by Drexler. This class includes (among other things) diamond crystals of arbitrary shape but with stably terminated surfaces (e.g., hydrogenated or the like) and with impurities at precise locations in the diamond lattice (e.g., substitutional boron). Our objective is to manufacture particular diamondoid structures once the location and type of every atom has been specified by design.

Interest of the Computer Industry

The attraction of molecular manufacturing for the computer industry should be clear. It should let us make computers at a manufacturing cost of less than a dollar per pound, operating at frequencies of tens of gigahertz or more, with linear dimensions for a single device of roughly 10 nanometers, high reliability, and energy dissipation (using conventional methods) of roughly 10^{-18} joules per logic operation. If we make thermodynamically reversible computers (which the author and others have recently shown can be made from conventional electronic devices, e.g., CMOS) then the energy dissipation per logic operation can be reduced to well below kT at T = 300 Kelvins (well below 10^{-21} joules). The computer industry is spending billions of dollars to make better computers.

It is widely acknowledged within the industry that lithography is approaching its limits. Articles like *The Future of the Transistor, Miniaturization of Electronics and its Limits* and *Outlook for VLSI: Will the Balloon Burst?* quite clearly show that conventional lithography will run out of steam (in perhaps a decade, though there is less agreement about the exact time frame).

There is already interest in molecular logic devices and that interest will increase sharply as improvements in conventional manufacturing methods become increasingly difficult. However, any new proposal for manufacturing molecular computers will be weighed against (at least) the criteria mentioned above.

If it cannot easily beat conventional methods after they have been pushed to their uttermost limits, then it will be rejected.

The computer industry will soon be pouring vast sums into research aimed at molecular computing, but the great bulk of funding will go towards well thought out proposals that offer a realistic possibility of substantially exceeding the performance of the ultimately evolved silicon VLSI technology that we expect to develop over the next decade. If you can't beat tomorrow's mainstream computers, you might as well not try.

The Problem

For this and many other reasons the class of diamondoid structures is a reasonable one to consider. The problem of building a diamondoid electronic computer captures many of the fundamental issues in molecular manufacturing, and poses clearly the issue of building large structures that cannot be made by regular repetition of some substructure (e.g., the unit cell of a crystal or the monomeric unit in a bio-polymer).

This brings us to a core issue in molecular manufacturing: how do we synthesize such things?

How We Make Diamond Today

Today, we can synthesize diamond at low pressure and low temperature by using CVD (Chemical Vapour Deposition) methods. Diamond CVD growth involves highly reactive species (radicals, carbenes, etc.) in a gas over the growing diamond surface that bombard and react with that surface at random. Because reaction sites are random, growth of many defect structures occurs (dislocations, etc.) as well as the desired perfect diamond structure.

Two fundamental mechanisms in the growth process include abstraction of hydrogens from the diamond surface leaving behind reactive sites (dangling bonds, radicals) and interaction of carbon species (both reactive (CH_2, CH_3, etc.) as well as relatively unreactive species (C_2H_2)) with the surface, thus depositing carbon.

If we are to synthesize diamondoid structures it is plausible that we begin our search for the basic reaction steps involved in this synthesis by looking at existing reactions that occur in the CVD growth of diamond. The use of a reactive gas in the synthesis process, however, would seem to defeat any hope of making precisely patterned diamondoid structures, for the gas will interact with the growing surface at random.

Positional Control is Fundamental

Here, we introduce the fundamental concept of molecular manufacturing: positional control over the site of reactions. To take a specific example we consider site specific hydrogen abstraction from the diamond surface. The ability to remove specific hydrogen atoms from the surface of the diamondoid work piece under construction is likely to be a fundamental unit operation in any attempt to make atomically precise diamondoid structures.

Hydrogen abstraction during CVD diamond growth typically involves a radical reaction between atomic H from the gas with H bonded to carbon on the surface producing H_2. It is unclear how to make this process site specific. However, there are other structures with a high affinity for hydrogen which offer greater possibilities for positional control. In particular, the propynyl radical C_3H_3 (figure) has a great affinity for hydrogen. Further, this radical has the very useful property that it has two ends: one end is a highly reactive radical while the other end is a stable sp^3 carbon. Thus, we could synthesize a larger molecule with the propynyl radical at its end. The larger molecule would be held at the tip of a positional device. The positional device would provide control over the orientation and position of this hydrogen abstraction tool (e.g., a six degrees of freedom manipulator) and thus control the site of abstraction by controlling the position of the tool.

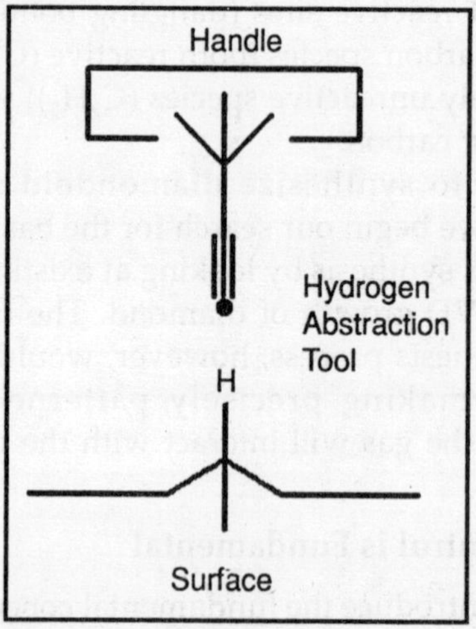

Fig. A Site Specific Hydrogen Abstraction Tool.

Ab initio quantum chemical analysis of the abstraction of hydrogen from isobutane using an ethynyl abstraction tool supports the idea that the barrier to this reaction is zero. The reaction will proceed rapidly and, because of the large exothermicity, irreversibly. Calculated barriers for abstraction from several other molecules were also small, suggesting that this hydrogen abstraction tool could be used to abstract hydrogen from a wide range of different molecules. Molecular dynamics simulations provide evidence that the abstraction reaction will select the correct hydrogen atom in the face of thermal noise at room temperature, as well as providing further support for the basic mechanism.

The site specific abstraction of hydrogen illustrates the core concept in molecular manufacturing: selecting the reaction site by controlling the position and orientation of the reactants. The (relatively stiff) diamondoid workpiece is held in place, while a tool (in our example, a hydrogen abstraction tool) is positioned using a rather conventional (if also rather small) robotic arm.

The ideas of using tools, controlling the position of those tools with a general purpose manipulator, and building complex structures by putting together components using those positionally controlled tools are rather common and even mundane at the macroscopic level.

At the molecular level, they are new and almost shocking: yet it is simply mapping onto the molecular world the concepts and ideas that have proven so useful and powerful in macroscopic manufacturing. By adding positional control we should be able to develop a method of molecular manufacturing which combines the best features of both conventional macroscopic manufacturing and chemical synthesis.

Other Molecular Tools

If we are to grow diamond, we must also have carbon deposition tools. Drexler has suggested the use of positionally controlled carbenes and alkynes and proposed reaction pathways and surface structures where these tools would apply.

In both cases, the tools are positioned at a precise point on the growing diamondoid structure and are used to deposit one or more carbon atoms at a desired location.

These deposition reactions parallel proposals in the CVD literature except for the addition of positional control (e.g., at least one portion of the moiety must be part of an extended "handle" which can be held by a positional device). These are only two examples from the wide range of tools that are capable of depositing carbon on a surface.

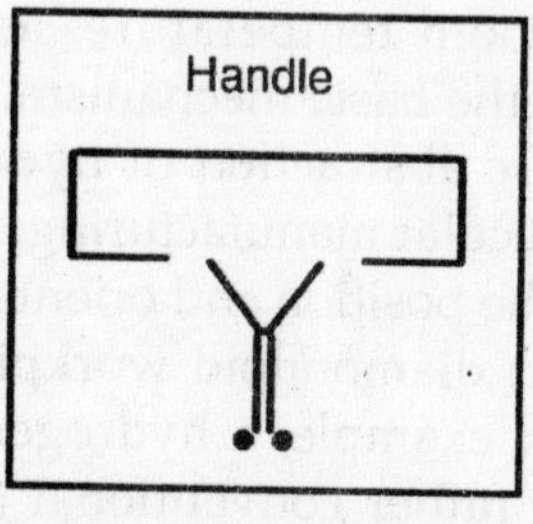

Fig. A Positionally Controlled Carbene

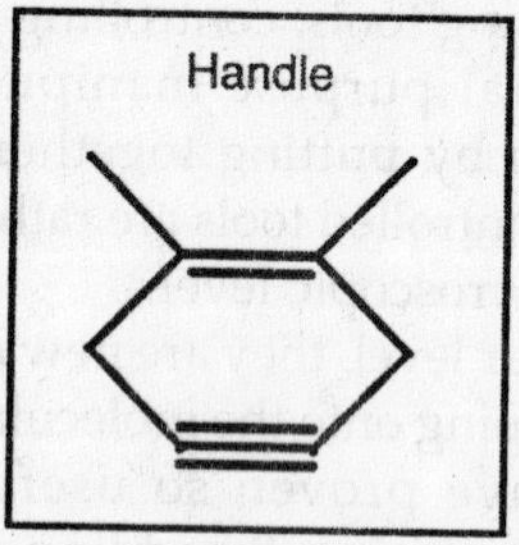

Fig. A Positional Controlled Strained Cycloalkyne

The broad range of possible tools coupled with the great power of ab initio computational chemistry should let us define and verify a complete set of molecular tools capable of synthesizing essentially any diamondoid structure. The work by Musgrave et. al. and Sinnott et. al. are first steps toward this objective.

Modern ab initio methods can produce results that are sufficiently accurate for this type of analysis. Further research in this area is feasible and should be pursued.

The Context of Tool Use

For such tools to be usable in a system context we must satisfy certain constraints. First and foremost, we must have a device capable of positioning the tool to within something like an atomic diameter.

On the diamond surface, the distance between adjacent hydrogens is about 2.5 Angstroms. Thus, positional accuracy of 1 to 2 angstroms for the hydrogen abstraction tool is required to prevent abstraction of the wrong hydrogen.

Second, because the tools can be highly reactive, we require an inert environment.

A simple inert environment is vacuum. Compressed helium or some other inert gas would also work. Third, because it is the relative tool-workpiece position that must be controlled, the workpiece under construction must be relatively rigid (e.g., not subject to vibrational motions that would exceed about an angstrom).

Fourth and last, we must have some way of generating the sometimes highly reactive tools (e.g., we need to define a precursor to the hydrogen abstraction tool, as well as precursors to the other tools).

In some sense, the analysis that we will now pursue is similar in type to retrosynthetic analysis. We start with the final product that we wish to build (a macroscopic diamondoid computer, for example), and then consider the possible predecessor structures which would yield the final product in one step.

Then we consider the predecessor structures to those structures, and so on. We extend retrosynthetic analysis beyond its traditional bounds, but the general concept remains the same: given the finished product we deduce the possible ways in which it could have been constructed. This approach has been called"backwards chaining"by Drexler.

Positional Devices and Molecular "Arms"

Work with SPMs (Scanning Probe Microscopes) clearly show that it is possible to achieve positional accuracies of a small fraction of an angstrom.

Small (~0.1 microns) diamondoid "arms" or positional devices with similar positional accuracy are in principle quite feasible. The field of robotics provides a broad range of designs for positional devices which are largely scale independent. Shrinking these designs to submicron size is conceptually straightforwards.

A factor of crucial importance in the design of molecular-scale positional devices is the accuracy with which the tip can be positioned, particularly in the face of thermal noise.

Whileatomically precise bearings and joints will not suffer from chatter, backlash, wear, tooth-to-tooth errors and other sources of inaccuracy caused by imprecise manufacturing, they will still suffer from positional errors caused by thermal noise. To control this source of error, it is essential that the robotic arm be very stiff, and so the use of stiff materials is desirable.

The Young's modulus for diamond is about 10^{12} Pascals (very stiff), and back-of-the-envelope calculations show that a hollow cylinder of such material that is perhaps 100 nanometers long and 30 nanometers in diameter should have a positional accuracy at the tip, in the face of thermal noise at room temperature, of a small fraction of an atomic diameter.

A more detailed design and analysis of a jointed tubular robotic arm taking into account the bending and rocking motions of joints in the arm further supports this conclusion. Alternatives to the simple robotic arm are available which might be more attractive.

Other Requirements for Tool Use

Creating an inert environment also presents no fundamental problems: high quality vacuums are common in Labouratories today. If our objective is to have a very small very high quality vacuum, then a relatively thin wall of diamondoid material could be used as a barrier to keep a volume which was a modest fraction of a cubic micron free of any contaminants.

If the volume were initially constructed free of contaminants then such a barrier would keep the inside free of any contaminants with high probability.

Because we are building diamondoid structures, they will be very stiff. As a consequence, it is relatively easy to meet the requirement that the objects that are being manufactured must themselves be stiff. Finally, generation of "activated" tools from relatively stable precursors can be done by a variety of methods.

Because we are assuming an environment in which we have positional control we can use particularly simple precursors. We illustrate this by considering a precursor to the hydrogen abstraction tool. This precursor has two handles, and X is chosen so that the X-C bond is weaker than the C-C bond. X might be Si or Ge.

If we pull on the two handles with sufficient force, something will break. Because the X-C bond was deliberately selected to be weaker than the other bonds in the structure, it will break. This gives us the activated hydrogen abstraction tool.

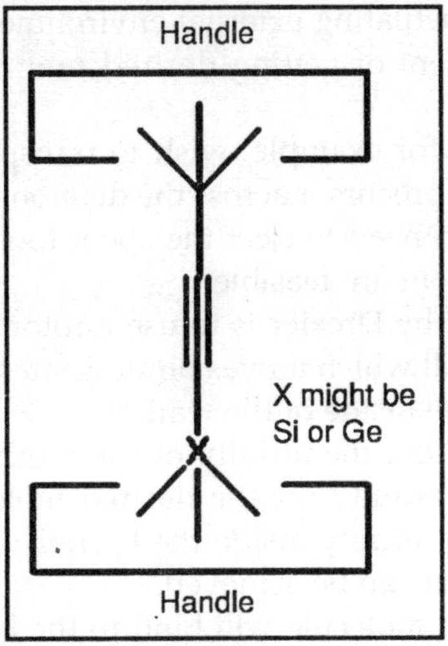

Fig. A Precursor to the Hydrogen Abstraction Tool.

A related question is: How can we get the hydrogen off the tip of the abstraction tool? A simple answer is: don't. Throw the tool away after one use. In a system design using this approach, it would be necessary to provide a continuous stream of precursors.

These would be activated, used once, and then discarded. A more elegant approach would be to remove the hydrogen from the tip and recycle the tool, as discussed by Drexler and Musgrave et. al.

More generally the activation of relatively stable precursors can be done by using any of several forms of energy: mechanical, optical, chemical or other. While the use of mechanical means to provide the activation energy for chemical reactions is relatively novel, in an environment where positional control is already available it is quite natural.

Selective Transport Across a Barrier

Having introduced a diamondoid barrier to keep unwanted contaminants out (much as the bacterial wall allows bacteria to maintain an appropriate internal environment in the face of a fluctuating external environment) we must now solve the problem of getting desired raw materials through the barrier.

We might, for example, wish to transport the hydrogen abstraction tool precursor across the diamondoid barrier. After use we will also need to eject the spent tool. Several ways to solve this problem are feasible.

A proposal by Drexler is to use a rotor embedded in the diamondoid wall which moves binding sites from the outside of the wall to the inside of the wall.

By modulating the affinity of the binding site so that it will have high affinity for the desired molecule outside the barrier and low affinity inside the barrier, efficient transport across the barrier can be achieved.

The desired molecule will bind to the binding site when it is outside the barrier, the binding site will be rotated to the inside of the barrier and the binding affinity reduced (in the illustrated proposal by mechanically pushing a rod into the binding site, thus physically precluding occupancy), and the molecule will be released on the inside of the barrier.

The result is to increase the concentration of the desired molecule. A few stages of such a filtration system can achieve extremely high purities.

The final stage, rather than ejecting the molecule into a liquid, would deliver the molecule into the inert internal environment in a well defined orientation where it could be further processed.

One simple method of further processing would be for the oriented molecule to be directly transferred to the tip of the positional device.

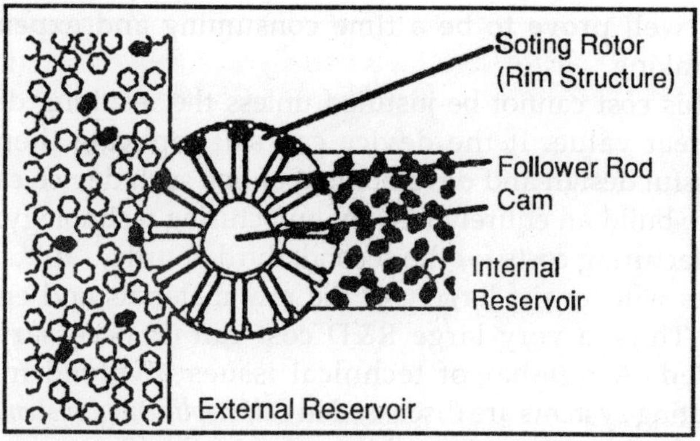

Fig. Selective Transport of Desired Molecules *From Nanosystems*: Molecular machinery, Manufacturing And Computation

Control Signals

Finally, we will need a source of control signals for our molecular arm. One general approach would be to use a molecular computer. We will not consider a particular design for a molecular computer here, it is sufficient to note that many proposals for molecular computation have been considered in the literature and it is generally expected that some type of very small computational device will be feasible in a few decades.

Summary

This completes our (all too brief) outline of a small device able to manufacture a broad class of diamondoid materials. Basically, the design is driven by the desire to provide the environment needed to synthesize diamond and diamondoid materials using the kinds of reactions that occur naturally during CVD growth of diamond.

The device is itself made from diamondoid materials, which means that one such device can manufacture a second such device.

This ability to self replicate is crucial in achieving low manufacturing costs. As the reader might appreciate, the design and construction of one such general purpose device might well prove to be a time consuming and expensive undertaking.

This cost cannot be justified unless the resulting device has great value. If the device can self replicate then the successful design and development of one such device can be used to build an entirely new manufacturing technology. The manufacturing costs for the second, third, fourth.... 10^{10}.... etc. devices will consist largely of the raw materials and energy costs. Thus, a very large R&D cost can (if necessary) be justified. A number of technical issues involved in self replicating systems are discussed in *Self Replicating Systems and Molecular Manufacturing* while some of the obvious safety issues are discussed in *The Risks of Nanotechnology*.

Bricks Without Vacuum

The major driving force in the previous design was the desire to approach fundamental limits (strength, stiffness, thermal conductivity, electrical characteristics, etc.) in the manufactured products. This in turn implied the use of diamondoid materials, which, when coupled with the known chemistry of CVD diamond synthesis resulted in a high vacuum system with highly reactive intermediates.

Relaxing the materials requirements gives us a *much* wider range of possible structures. In particular, we can consider what is sometimes called "brick" or "building block" based nanotechnology. In this approach we first design a set of molecular building blocks, and then assemble the building blocks by the use of positional control. The ribosome can be viewed as the prototype for this approach. The building blocks are amino acids, and they are linked together by the ribosome to form proteins.

Our approach differs in two principle respects: first, we add positional and orientational control over the building blocks in three dimensions, while the ribosome can only build structures that are fundamentally one dimensional (relying on linear structures that spontaneously fold into a particular shape to achieve a degree of control in three dimensions). Second, rather than using relatively floppy polymers, we prefer relatively rigid bricks that can be bonded to each other in a stiff three-dimensional framework.

In general, molecular structures built of bricks will be:

- Larger
- Weaker
- Less stiff
- Have poorer thermal conductivity
- Have poorer electrical properties
- Etc. etc. etc.

The sacrifice made in materials properties is significant. On the other hand, many bricks can be assembled in more conventional environments (solution), and so we can eliminate the need for vacuum. This greatly simplifies the system. Indeed, with brick-based nanotechnology one can relatively easily envision the synthesis of a set of bricks that can, with the addition of positional control, be assembled into a wide range of structures with the stiffness of wood.

Molecular Manipulators

A conceptually simple and relatively near-term way to achieve some degree of positional control would be to use conventional SPMs. The tips used in current SPM's are usually quite crude, and even when it is possible to make a very fine tip the range of possible structures is very limited (tungsten or some other simple material is generally all that's available).

The design of tips for the SPM that incorporate individual molecules specifically synthesized for the purpose is a likely next step, and one that seems essential if we are to make progress in using SPM's to guide chemical reactions in a selective way. Such a "molecular manipulator" should be within reach of today's experimental technology. While the Young's modulus of the things it could make would be substantially inferior to that of diamond, this can be compensated by making them bigger. Scaling laws are such that increasing the size of an object by a factor of 10 also increases its stiffness by a factor of 10, and so reduces the positional inaccuracy from thermal vibration by a factor of 10.

If we wish to build a molecular "arm" out of bricks, then

to achieve the same positional accuracy as with a diamondoid arm we will have to make the arm bigger — perhaps several tenths of a micron or more — but the basic design concepts that were discussed previously for use with a diamondoid arm still hold.

We can still position the individual bricks accurately both in position and orientation, we can still build larger structures by putting together many bricks, we can still control the synthesis process by positional means, we can still make a positional device from bricks, etc.

Why pursue such an approach when it can only make relatively inferior materials?

There are three primary reasons:

- It's easier to do.
- It could still make many things that are very valuable By today's standards.
- Such systems could be used to make better systems (e.g., diamondoid systems).

As the reader will appreciate, there are many possible candidates for bricks. Many researchers are already considering the design of molecular building blocks, although in most cases they do not consider positional control.

Krummenacker discusses several of the issues surrounding the design of molecular bricks intended for assembly via positional control. Adding positional control makes the synthesis of a broad range of molecular structures feasible, but at the same time requires the design and synthesis of a mechanism able to provide such control.

The obvious first step is to provide positional control using relatively modest extensions to today's SPMs. The most significant addition is the inclusion of a molecular tip, for today's SPMs typically have a limited range of possible tip structures and are often imprecise at the molecular scale.

MASSIVE PARALLELISM

Nanotechnology can make big things as well as small things. An attractive approach is to use *convergent assembly*, which can rapidly make products whose size is measured in meters starting from building blocks whose size is measured in nanometers. It is based on the idea that smaller parts can be assembled into larger parts, larger parts can be assembled into still larger parts, and so forth. This process can be systematically repeated in a hierarchical fashion, creating an architecture able to span the size range from the molecular to the macroscopic. Convergent assembly covers a class of many different architectures. Drexler described one member of this class . The present proposal is simpler while retaining the desirable performance characteristics described in *Nanosystems*

Convergent Assembly

It's common to make bigger things by assembling them from smaller things. A finished assembly which is about 1 meter in size might be made from eight smaller subassemblies each of about 0.5 meters in size. To make a 1 meter cube, we'd need to assemble eight smaller cubes, each of 0.5 meters in size. This relationship holds approximately for many products.

We could assemble our eight subassemblies into a finished assembly by using one or more robotic arms (or other positional devices) in an assembly module (depicted abstractly in figure). The present paper is focused on architectural issues; the reader interested in details of how positional devices can be used in manufacturing is referred to other sources – one example is *A new family of six degree of freedom positional devices* (Merkle, submitted to *Nanotechnology*).

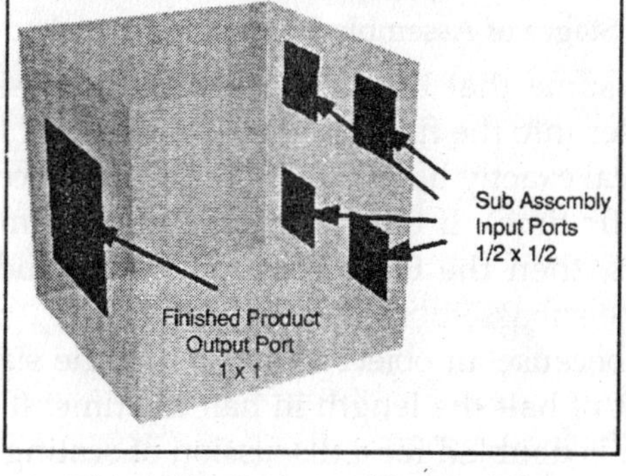

Fig. A Single Assembly Module.

Which produces a finished assembly to the left through the single output port. If the output port has a size of one meter by one meter, each input port would have a size of 0.5 meters by 0.5 meters.

This process could be repeated. We might take 64 sub-subassemblies, each with a size of about 0.25 meters, assemble them into 8 subassemblies, each with a size of about 0.5 meters, and finally assemble the 8 subassemblies into the finished assembly, with a size of about 1.0 meters.

This could be done using the system shown in figure , which depicts a 1.0 meter assembly module connected to four 0.5 meter assembly modules. (Note that we are designating an "assembly module" by the size of its output port. The module size is larger — in this case, the 1.0 meter assembly module has a size of 2.0 meters. This additional size provides sufficient room to handle subassemblies and the final 1.0 meter assembly).

Fig. Two Stages of Assembly Producing a Finished Product.

Let's assume that the final stage (which assembles the subassemblies into the final product) takes about 100 seconds. (Its not critical exactly how long it takes, but it's convenient to pick a specific time). If the 1.0 meter assembly module takes 100 seconds, then the 0.5 meter assembly module should complete one subassembly in 50 seconds.

This is because an object which is half the size can finish a movement of half the length in half the time: it's frequency of operation is doubled for a discussion of scaling laws). Each 0.5 meter assembly module can therefore produce two subassemblies in 100 seconds. The four 0.5 meter assembly modules can finish eight subassemblies in 100 seconds, which means that both the 1.0 meter assembly module and the four 0.5 meter assembly modules must work for 100 seconds to produce the final product. This process can, of course, be continued.

Figure shows three stages of this process, in which 512 sub-sub-subassemblies are assembled in 16 0.25 meter assembly modules, making 64 sub-subassemblies; these 64 sub-subassemblies are assembled into 8 subassemblies in 4 0.5 meter assembly modules. Finally, the 8 subassemblies are assembled into the final product in the single 1.0 meter assembly module.

Fig. Three Stages of Convergent Assembly.

Again, the 0.25 meter assembly modules operate twice as fast as the 0.5 meter assembly modules, and four times as fast as the 1.0 meter assembly module. The 16 0.25 meter assembly modules can make 64 sub-subassemblies in the same time that the 4 0.5 meter assembly modules can make 8 subassemblies, which is the same amount of time that the single 1.0 meter assembly module takes to produce the finished product. We can keep adding stages to this process until the inputs to the first stage are as small as we might find convenient. For the purposes of molecular manufacturing, it is convenient to assume that these initial inputs are ~one nanometer in size.

Summary of Convergent Assembly

The following rules hold approximately for a convergent

assembly system which doubles the size of product produced at each stage, which starts with building blocks with a characteristic length of λ meters, and which has a characteristic assembly time to manipulate those building blocks of τ seconds. Values for λ and τ might be 10^{-9} meters (one nanometer) and 10^{-7} seconds (100 nanoseconds).

- The size of product that can be made at stage N is $2^N \times \lambda$ meters.
- The time required for an assembly module in the Nth stage to produce an assembly of size $2^N \times \lambda$ meters given that eight subassemblies of size $2^{(N-1)} \times \lambda$ meters are already available is $2^N \times \tau$ seconds. We will further assume that the assembly process can be started when only four subassemblies are available, effectively allowing stage N to begin operation after stage N–1 has produced only half the required subassemblies.
- The time required to produce a product of size $2^N \times \lambda$ meters, including the time taken by all preceding stages (i.e., assuming all stages start empty, with no assemblies or subassemblies in them, but that they have been provided with appropriate instructions for the manufacture of the desired product) is $2^{(N+1)} \times \tau$ seconds.
- Such a system can be pipelined, in which case a steady stream of finished products of size $2^N \times \lambda$ meters can be produced at a rate of one new product every $2^N \times \tau$ seconds.
- The area of the output port of an assembly module equals the sum of the areas of its input ports. The total area of all output ports at stage N is equal to the total area of all output ports at stage N–1.
- The average speed with which material flows through the system is approximately equal at all stages, and is approximately λ / τ meters per second. For the example values of λ and τ chosen here, this speed is roughly 0.01 meters/second.

The example values selected for τ and λ imply that the assembly of a 1.0 meter product from eight 0.5 meter subassemblies will take about 100 seconds. There seems to be no particular reason to assume that the characteristic time τ could not be made shorter, thus increasing the speed of the manufacturing process.

The claim that it requires $2^{(N+1)} \times \tau$ seconds for the Nth stage to produce a single product of size $2^N \times \lambda$ meters, made in item 3 above, can be shown by induction. We simply postulate that stage 1 requires $4 \times \tau$ seconds, thus providing the base for the induction. Assume that the hypothesis is true for stage N-1 and we wish to show that it is true for stage N.

The assembly modules at stage N-1 will produce their first output in $2^N \times \tau$ seconds, at which point (by item 2) the assembly process can begin in stage N. The assembly process in stage N adds an additional $2^N \times \tau$ seconds (by item 2) before it produces its first output, so the total time before stage N produces an output is $(2^N + 2^N) \times \tau$ seconds, or $2^{(N+1)} \times \tau$ seconds, as desired.

If we assume that stage N is unable to overlap any operations at all with stage N–1, i.e., we assume that stage N-1 must produce all eight subassemblies before stage N can even begin, then we will increase the time needed before the Nth stage can produce its first output to $3 \times 2^N \times \tau$ seconds. This is only 50% longer.

A 30 stage system (N=30) should be able to produce a single meter-sized product in ~200 seconds. It should be able to produce a steady stream of meter-sized products at a rate of roughly one new product every 100 seconds.

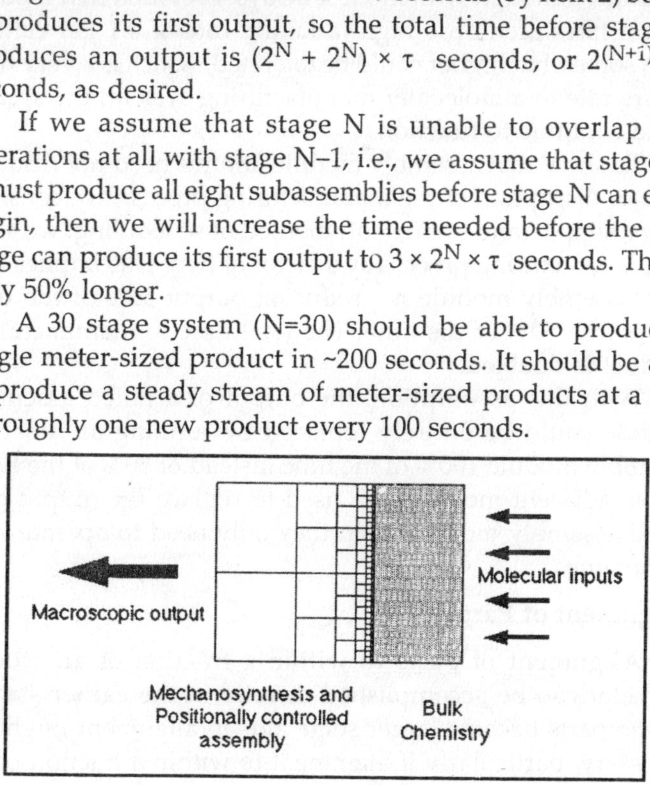

Assembly Module Failures

If an assembly module fails (which we assume can be detected, if not immediately then by the subsequent assembly module), then other assembly modules can take over the task of producing the output which should have been produced by the failed module.

As each assembly module accepts parts from four subassembly modules, the failure of a single subassembly module implies that the assembly module will receive parts from only three subassembly modules.

Provided that at least one of the three surviving subassembly modules can be reprogrammemed to build the subassemblies that would have been produced by the failed subassembly module, then the manufacturing process can continue, although with modest delays. Provided that module failure rates aren't too high (a failure rate of 0.1 Per cent or even somewhat higher could be tolerated, which is a very high failure rate in a molecular manufacturing system) the system could continue to function.

To show that assembly module failures need not result in a major disruption of the manufacturing process, we consider a very simple (and inefficient) method of scheduling: we slow the manufacturing process by a factor of two. This implies that each assembly module is producing output 50% of the time and is idle 50% of the time. The total time to manufacture a product is doubled.

With this assumption, the output of a failed assembly module could be entirely replaced by running an adjacent assembly module 100% of the time instead of 50% of the time. If two adjacent modules are used to replace the output of a failed assembly module, then they only need to operate 75% of the time.

Alignment of Parts

Alignment of parts to within a fraction of an atomic diameter can be accomplished directly in the earlier stages. As the parts become larger some aids in alignment might be necessary, particularly if alignment to within a fraction of an atomic diameter is desired. Several approaches for doing this are feasible.

One particularly straightforward approach (discussed in *Nanosystems*, is to use alignment pegs. If two surfaces are to be joined, one surface might have protruding pegs while the other would have corresponding holes.

By tapering both the end of the peg and the start of the hole, even if the two surfaces are initially misaligned, the peg will enter the hole.

As the two surfaces approach more closely, they will be forced into alignment by the peg and hole. Precise alignment will be particularly important if wires or other fine structures must be precisely joined. A variety of alignment methods are feasible.

Accurate alignment of parts in vacuum should also allow reactive surfaces to be bonded together with great precision. While a wide range of possibile surfaces exist, one surface that might be worth further investigation is the diamond surface. This surface does not reconstruct, and so bringing together two diamond surfaces in vacuum might be an effective method of directly bonding two diamond blocks to each other.

Modeling of the energy released during this process and its possible adverse effects on the diamond lattice in the vicinity of the plane joining the two blocks would be useful. If the energy released during bonding causes significant damage, inclining the two surfaces with respect to each other at a small angle and then joining them slowly should create a more controlled rele·se of energy at the line of joining.

The First Few Stages

The analysis given here is based on simple scaling laws. Various issues influence the validity of the scaling laws in the first few stages.

At the smallest size molecular "parts" might well be found in solution. It will be necessary to bind and orient suitable feedstock molecules at a speed commensurate with the needs of the rest of the system if the indicated speeds of manufacturing are to be achieved.

Nanosystems proposed the use of variable affinity binding sites to orient molecules from the feedstock and bring them into the system. High concentrations of feedstock molecules could be employed to insure that this process takes place rapidly. In the extreme case, the "solvent" could itself be the feedstock molecule used in further assembly steps.

More generally, if the speed of operation of the binding sites is insufficient to provide feedstock molecules at a sufficient rate, then more binding sites can be employed. The area exposed to the feedstock solution can be increased by using relatively straightforward methods (e.g., multiple thin sheets which have binding sites on their surfaces, etc). It is not expected that this will pose a fundamental problem.

Another issue that arises in the first few stages is the scaling of the control system. At the macroscopic scale, we are used to the idea that a computer is much smaller than the robotic arm which it controls. For the very smallest scales, this assumption is dubious.

A simple 8-bit processor might occupy a cube about 100 nanometers on a side. If we assume the linear dimensions of each assembly module are only a few times larger than the size of the components to be assembled, then our 8-bit processor will dominate the volume requirements of the assembly module for the first few stages. If the first stage handles one nanometer parts, then our simple scaling laws suggest that the first stage assembly module will have linear dimensions perhaps three times that size (e.g., 3 nanometers). As our 8-bit control processor is over 30 times larger than this (in linear dimensions) the validity of our scaling laws is evidently open to serious question.

One approach to this problem would be to "special case" the first few stages. By employing hard automation and eliminating their flexibility, the control requirements could be substantially reduced or effectively eliminated. This seems like a good idea, as the advantages of hard automation over more flexible methods should be substantial. By combining a relatively modest number of standard 100 nanometer parts we could make a remarkably wide range of products.

If, however, we want a manufacturing process which can more closely approach the ultimate limit of being able to manufacture almost any product consistent with the laws of chemistry and physics, then we could adopt a more flexible approach: we could allow the first few stages to deviate from the simple scaling laws used for the rest of the stages. This is feasible because the first few stages occupy very little volume: if we continued to follow the same scaling laws as used for the later stages, the first few stages would be little more than a very thin "skin" at the beginning of the assembly process. We can deviate from the scaling laws adopted for larger stages by increasing the thickness of this "skin."

We could eliminate the first few stages and replace them with special assembly modules that:

- Accept as inputs parts of about 1 nanometer but
- Produce as output parts of about 100 nanometers in size.

This means that the volume occupied by the control system is roughly similar in volume to the rest of the special assembly module. Of course, this means that the first 6 to 7 "normal" stages have been eliminated. If we simply had one layer of special assembly modules, we would find that they were too slow to feed the rest of the system. As suggested above, we can increase the number of special assembly modules and let them occupy a larger volume.

We have been assuming that each assembly module would assemble about 8 parts while operating at a certain characteristic speed. Our special assembly modules will operate at the characteristic speed of stage 6 or 7, but will be handling $(2^6)^{3/8} = 32{,}000$ to $(2^7)^{3/8} = 262{,}000$ times as many parts. Another way of phrasing this is to point out that a cube 100 nanometers on a side is composed of 1,000,000 one-nanometer cubes.

Thus, a cube 100 nanometers on a side has about 1,000,000 one-nanometer parts. As a consquence, our special assembly module will produce output parts at a speed roughly 100,000 times slower than it would produce output parts which had only 8 sub-parts in them. This implies that we need roughly

100,000 times more special assembly modules to produce parts at the same rate as would otherwise have been produced by the 7th stage modules if we had kept using our original scaling laws. As the 6th or 7th stage modules are a few hundred nanometers in size, we must increase the thickness of the first 6 to 7 stages to a few centimeters (100,000 times a few hundred nanometers). It should be emphasized that this increase in volume is based on the assumption that we wish to retain the greatest possible flexibility in control over the first few stages by using one computer per assembly module.

Limitations

Convergent assembly is likely to prove most useful in the manufacture of relatively rigid structures made from rigid components. This should include computers with a large number of molecular logic gates densely packed into a three dimensional volume; structural elements useful in load bearing applications in cars, airplanes, bicycles, buildings, furniture, etc.; high density memories; and many more. In aerospace applications the great strength-to-weight ratio of diamond (over 50 times that of steel) should prove particularly useful.

Although each stage in our example system was a factor of two larger than the preceding stage, it is not clear that a factor of two is the most appropriate value for all (or even most) manufacturing applications: larger factors might prove more convenient for some operations. If we wish to bolt two parts together, the bolt might be an order of magnitude smaller than either part.

This can be accomodated by passing the bolt through several stages of binary convergent assembly and finally using it in a stage which can accomodate much larger parts. More generally, small parts can, if necessary, be passed through several stages unchanged if this should prove convenient or necessary.

The inefficiencies of this approach might be offset by making many small parts and passing them through succeeding stages in a batch (e.g., one might make a box with many bolts in it, rather than making a single bolt and passing it alone through several successively larger and increasingly underutilized stages).

Alternatively, it might be advantageous to increase the size of the succeeding stage by a factor larger than two. Complex patterns in which different stages or even different assembly modules in each stage increase in size by individually selected amounts might prove useful.

The greatest limitation of convergent assembly as presented here is the requirement that the product be relatively stiff. Easily deformed, liquid or gaseous components are most problematic.

An inflated balloon (with thanks to Josh Hall for suggesting this example) does not appear well suited to convergent assembly. Many products, however, could be initially manufactured in a more rigid form (e.g., at lower temperature or with supports or scaffolding to provide the desired rigidity) and then later allowed to assume the desired less rigid state (by removing the scaffolding, warming, etc).

Other Approaches

We have considered a relatively straighforward system in which parts move only in one direction. Parts could also move laterally and backwards, which might be useful in bypassing failed modules.

Convergent assembly is only one approach to making large objects from small objects. An oak tree does not use convergent assembly, but quite effectively makes something large starting with only a small seed. We have considered an approach in which relatively rigid"parts"have approximately similar X, Y and Z sizes.

It would be possible to consider"parts"which are basically one dimensional. This might be useful in the manufacture of products similar to cables, in which long thin strands are woven together into a final product.

Two dimensional"parts,"e.g., sheets, could also be produced and then pressed together into a final product. Clearly, we would be happy with any method that simultaneously achieved the first three objectives.

However, this seems difficult without using some form of positional assembly (to get the right molecular parts in the right places) and some form of massive parallelism (to keep the costs down). The need for positional assembly implies an interest in molecular robotics, e.g., robotic devices that are molecular both in their size and precision. These molecular scale positional devices are likely to resemble very small versions of their everyday macroscopic counterparts. Positional assembly is frequently used in normal macroscopic manufacturing today, and provides tremendous advantages.

Imagine trying to build a bicycle with both hands tied behind your back! The idea of manipulating and positioning individual atoms and molecules is still new and takes some getting used to. However, as Feynman said in a classic talk in 1959:"The principles of physics, as far as I can see, do not speak against the possibility of maneuvering things atom by atom.

"We need to apply at the molecular scale the concept that has demonstrated its effectiveness at the macroscopic scale: making parts go where we want by putting them where we want! One robotic arm assembling molecular parts is going to take a long time to assemble anything large - so we need lots of robotic arms: this is what we mean by massive parallelism.

While earlier proposals achieved massive parallelism through self replication, today's"best guess"is that future molecular manufacturing systems will use some form of convergent assembly.

In this process vast numbers of small parts are assembled by vast numbers of small robotic arms into larger parts, those larger parts are assembled by larger robotic arms into still larger parts, and so forth. If the size of the parts doubles at each iteration, we can go from one nanometer parts (a few atoms in size) to one meter parts (almost as big as a person) in only 30 steps.

APPLICATIONS OF NANOTECHNOLOGY

Nanotechnology Applications

The Understanding Nanotechnology Website is dedicated

to providing clear and concise explanations of nanotechnology applications. Scan the listings below to find an application of interest, or click any topic on the navigation bar to the left to go directly to the page discussing an application of interest.

Nanotechnology Applications in Medicine

Researchers are developing Coustomised nanoparticles the size of molecules that can deliver drugs directly to diseased cells in your body. When it's perfected, this method should greatly reduce the damage treatment such as chemotherapy does to a patient's healthy cells.

Nanotechnology in Medicine– Nanomedicine

The use of nanotechnology in medicine offers some exciting possibilities. Some techniques are only imagined, while others are at various stages of testing, or actually being used today.

While some researchers use the term nanomedicine to refer to applications of nanoparticles currently under development, other researchers reserve the term nanomedicine to refer to longer range research that involves the use of manufactured nano-robots to make repairs at the cellular level. Whatever you call it, the use of nanotechnology in the field of medicine could revolutionize the way we detect and treat damage to the human body and disease in the future, and many techniques only imagined a few years ago are making remarkable progress towards becoming realities.

Nanomedicine Application Area: Drug Delivery

One application of nanotechnology in medicine currently being developed involves employing nanoparticles to deliver drugs, heat, light or other substances to specific types of cells (such as cancer cells).

Particles are engineered so that they are attracted to diseased cells, which allows direct treatment of those cells. This technique reduces damage to healthy cells in the body and allows for earlier detection of disease.

For example, nanoparticles that deliver chemotherapy drugs directly to cancer cells are under development. Tests are in progress for targeted delivery of chemotherapy drugs and their final approval for their use with cancer patients is pending, as explained on CytImmune Science's website. CytImmune has published the preliminary results of a Phase 1 Clinical Trial of their first targeted chemotherapy drug.

If you hate getting shots, you'll be glad to hear that oral administration of drugs that currently are delivered by injection may be possible in many cases.

The drug is encapsulated in a nanoparticle which helps it pass through the stomach to deliver the drug into the bloodstream.

There are efforts underway to develop oral administration of several different drugs using a variety of nanoparticles. A company which has progressed to the clinical testing stage with a drug for treating systemic fungal diseases is BioDelivery Sciences, which is using a nanoparticle called a cochleate. You can read the initial results from their Phase 1 Clinical Study here.

Nanomedicine Application Area: Therapy Techniques

Buckyballs may be used to trap free radicals generated during an allergic reaction and block the inflammation that results from an allergic reaction. Nanoshells may be used to concentrate the heat from infrared light to destroy cancer cells with minimal damage to surrounding healthy cells.

Nanospectra Biosciences has developed such a treatment using nanoshells illuminated by an infrared laser that has been approved for a pilot trial with human patients. Nanoparticles, when activated by x-rays, that generate electrons that cause the destruction of cancer cells to which they have attached themselves. This is intended to be used in place radiation therapy with much less damage to healthy tissue. Nanobiotix has released preclinical results for this technique.

Aluminosilicate nanoparticles can more quickly reduce bleeding in trauma patients by absorbing water, causing blood in a wound to clot quickly. Z-Medica is producing a medical gauze that uses aluminosilicate nanoparticles.

Nanomedicine Application

Quantum Dots (qdots) may be used in the future for locating cancer tumors in patients and in the near term for performing diagnostic tests in samples.

Invitrogen's website provides information about qdots that are available for both uses, although at this time the use "in vivo" (in a living creature) is limited to experiments with lab animals. Iron oxide nanoparticles can used to improve MRI images of cancer tumors.

The nanoparticle is coated with a peptide that binds to a cancer tumor, once the nanoparticles are attached to the tumor the magnetic property of the iron oxide enhances the images from the Magnetic Resonance Imagining scan. Nanoparticles can attach to proteins or other molecules, allowing detection of disease indicators in a lab sample at a very early stage.

There are several efforts to develop nanoparticle disease detection systems underway. One system being developed by Nanosphere, Inc. uses gold nanoparticles, Nanosphere has clinical study results with their Verigene system involving it's ability to detect four different nucleic acids, while another system being developed by T2 Biosystems uses magnetic nanoparticles to identify specimens, including proteins, nucleic acids, and other materials.

Anti-Microbial Techniques

One of the earliest nanomedicine applications was the use of nanocrystalline silver which is as an antimicrobial agent for the treatment of wounds, as discussed on the Nucryst Pharmaceuticals Corporation website. A welcome idea in the early study stages is the elimination of bacterial infections in a patient within minutes, instead of delivering treatment with antibiotics over a period of weeks. You can read about design analysis for the antimicrobial nanorobot used in such treatments in the following article: Microbivores: Artifical Mechanical Phagocytes using Digest and Discharge Protocol.

Applications in Cell Repair

Nanorobots could actually be programmemed to repair specific diseased cells, functioning in a similar way to antibodies in our natural healing processes. Read about design analysis for one such cell repair nanorobot in this article: The Ideal Gene Delivery Vector: Chromallocytes, Cell Repair Nanorobots for Chromosome Repair Therapy

Nanotechnology in Medicine: Company Directory

Company	Product
BioDelivery Sciences	Oral drug delivery of drugs encapuslated in a nanocrystalline structure called a cochleate
CytImmune	Gold nanoparticles for targeted delivery of drugs to tumors
Invitrogen	Qdots for medical imaging
Nucryst	Antimicrobial wound dressings using silver nanocrystals
Luna Inovations	Bucky balls to block inflammation by trapping free radicals
NanoBio	Nanoemulsions for nasal delivery to fight viruses (such as the flu and colus) or through the skin to fight bacteria
NanoBioMagnetics	Magnetically responsive nanoparticles for targeted drug delivery and other applications
Nanobiotix	Nanoparticles that target tumor cells, when irradiated by xrays the nanoparticles generate electrons which cause localized destruction of the tumor cells.
Nanospectra	AuroShell particles (nanoshells) for thermal destruction of cancer tissue
Nanosphere	Diagnostic testing using gold nanoparticles to detect low levels of

	proteins indicating particular diseases
Nanotherapeutics	Nanoparticles for improving the performance of drug delivery by oral or nasal methods
Oxonica	Diagnostic testing using gold nanoparticles (biomarkers)
T2 Biosystems	Diagnostic testing using magnetic nanoparticles
Z-Medica	Medical gauze containing aluminosilicate nanoparticles which help blood clot faster in open wounds.
Sirnaomics	Nanoparticle enhanced techniques for delivery of siRNA

Nanotechnology in Medicine: Resources

National Cancer Institute Alliance for Nanotechnology in Cancer; This alliance includes a Nanotechnology Characterization Lab as well as eight Centres of Cancer Nanotechnology Excellence. Alliance for NanoHealth; This alliance includes eight research institutions performing colLabourative research.

European Nanomedicine platform: The National Institute of Health (NIH) is funding research at eight Nanomedicine Development Centres.

ELECTRONICS

Nanotechnology holds some answers for how we might increase the capabilities of electronics devices while we reduce their weight and power consumption.

Nanotechnology in Electronics (Nanoelectronics)

How can nanotechnology improve the capabilities of electronic components?

Nanoelectronics holds some answers for how we might increase the capabilities of electronics devices while we reduce their weight and power consumption. Some of the nanoelectronics areas under development, which you can explore in more detail by following the links provided in the next section, include the following topics. Improving display screens on electronics devices. This involves reducing power consumption while decreasing the weight and thickness of the screens.

Increasing the density of memory chips. Researchers are developing a type of memory chip with a projected density of one terabyte of memory per square inch or greater. Reducing the size of transistors used in integrated circuits. One researcher believes it may be possible to "put the power of all of today's present computers in the palm of your hand".

Nanoelectronics: Applications under Development

Researchers are looking into the following nanoelectronics projects: Building transistors from carbon nanotubes to enable minimum transistor dimensions of a few nanometers and developing techniques to manufacture integrated circuits built with nanotube transistors.

Using electrodes made from nanowires that would enable flat panel displays to be flexible as well as thinner than current flat panel displays. Using MEMS techniques to control an array of probes whose tips have a radius of a few nanometers. These probes are used to write and read data onto a polymer film, with the aim of producing memory chips with a density of one terabyte per square inch or greater.

Combining gold nanoparticles with organic molecules to create a transistor known as a NOMFET (Nanoparticle Organic Memory Field-Effect Transistor). Using carbon nanotubes to direct electrons to illuminate pixels, resulting in a lightweight, millimeter thick "nanoemmissive" display panel.

Making integrated circuits with features that can be measured in nanometers (nm), such as the process that allows the production of integrated circuits with 45 nm wide transistor gates.

Using nanosized magnetic rings to make Magnetoresistive Random Access Memory (MRAM) which research has indicated may allow memory density of 400 GB per square inch. Developing molecular-sized transistors which may allow us to shrink the width of transistor gates to approximately one nm which will significantly increase transistor density in integrated circuits.

SPACE

Nanotechnology may hold the key to making space-flight more practical. Advancements in nanomaterials make lightweight solar sails and a cable for the space elevator possible. By significantly reducing the amount of rocket fuel required, these advances could lower the cost of reaching orbit and traveling in space.

Nanotechnology and Spaceflight

Nanotechnology may hold the key to making spaceflight more practical. Advancements in nanomaterials make lightweight solar sails and a cable for the space elevator possible. By significantly reducing the amount of rocket fuel required, these advances could lower the cost of reaching orbit and traveling in space.

In addition, new materials combined with nanosensors and nanorobots could improve the performance of spaceships, spacesuits, and the equipment used to explore planets and moons, making nanotechnology an important part of the 'final frontier.'

Spaceflight and Nanotechnology: Applications Under Development

Researchers are looking into the following applications of nanotechnology in spaceflight: Using carbon nanotubes to make the cable needed for the space elevator, a system which could significantly reduce the cost of sending material into orbit. Nova has a nice video explaining the concepts.

Including layers of bio-nano robots in spacesuits. The outer layer of bio-nano robots would respond to damages to the spacesuit, for example to seal up punctures.

An inner layer of bio-nano robots could respond if the astronaut was in trouble, for example by providing drugs in a medical emergency.

Deploying a network of nanosensors to search large areas

of planets such as Mars for traces of water or other chemicals. Employing materials made from carbon nanotubes to reduce the weight of spaceships while retaining or even increasing the structural strength.

Producing thrusters for spacecraft that use MEMS devices to accelerate nanoparticles. This should reduce the weight and complexity of thruster systems used for interplanetary missions.

One cost-saving feature of these type of thrusters is their ability to draw on more or less of the MEMS devices depending upon the size and thrust requirement of the spacecraft, rather than designing and building different engines for different size spacecraft. Using carbon nanotubes to build lightweight solar sails that use the pressure of light from the sun reflecting on the mirror-like solar cell to propel a spacecraft. This solves the problem of having to lift enough fuel into orbit to power spacecraft during interplanetary missions. Working with nanosensors to monitor the levels of trace chemicals in spacecraft to monitor the performance of life support systems.

Spaceflight and Nanotechnology: Research Organizations

The Centre for Nanotechnology at NASA Ames is looking at how nanotechnology can be used to reduce the mass, volume, and power consumption of a wide range of spacecraft systems including sensors, communications, navigation, and propulsion systems. The Johnson Space Centre Nano Materials Project is working on nanotube composites with the aim of reducing spacecraft weight. The LiftPort Group is dedicated to making the space elevator reality. Their target date is October, 2031. The Space Nano-technology Labouratory at MIT is developing high perfor-mance instrumentation for use on spaceflights.

FOOD

Nanotechnology is having an impact on several aspects of food science, from how food is grown to how it is packaged. Companies are developing nanomaterials that will make a difference not only in the taste of food, but also in food safety, and the health benefits that food delivers.

How is Nanotechnology Being used in Food Science

Nanotechnology is having an impact on several aspects of food science, from how food is grown to how it is packaged. Companies are developing nanomaterials that will make a difference not only in the taste of food, but also in food safety, and the health benefits that food delivers.

Food Science: Current Nanotechnology Applications

Clay nanocomposites are being used to provide an impermeable barrier to gasses such as oxygen or carbon dioxide in lightweight bottles, cartons and packaging films.

Storage bins are being produced with silver nanoparticles embedded in the plastic. The silver nanoparticles kill bacteria from any material that was previously stored in the bins, minimizing health risks from harmful bacteria.

Food Science: Nanotechnology Applications under Development

Nanoparticles are being developed that will deliver vitamins or other nutrients in food and beverages without affecting the taste or appearance. These nanoparticles actually encapsulate the nutrients and carry them through the stomach into the bloodstream.

Researchers are using silicate nanoparticles to provide a barrier to gasses (for example oxygen), or moisture in a plastic film used for packaging.

This could reduce the possibly of food spoiling or drying out. Zinc oxide nanoparticles can be incorporated into plastic packaging to block UV rays and provide anti bacterial protection, while improving the strength and stability of the plastic film.

Nanosensors are being developed that can detect bacteria and other contaminates, such as salmonella, at a packaging plant. This will allow for frequent testing at a much lower cost than sending samples to a lab for analysis. This point-of-packaging testing, if conducted properly, has the potential to dramatically reduce the chance of contaminated food reaching grocery store shelves.

Research is also being conducted to develop nanocapsules containing nutrients that would be released when nanosensors detect a vitamin deficiency in your body. Basically this research could result in a super vitamin storage system in your body that delivers the nutrients you need, when you need them.

"Interactive" foods are being developed by Kraft that would allow you to choose the desired flavour and Colour. Nanocapsules that contain flavour or Colour enhancers are embedded in the food; inert until a hungry consumer triggers them. The method hasn't been published, so it will be interesting to see how this particular trick is accomplished.

Researchers are also working on pesticides encapsulated in nanoparticles; that only release pesticide within an insect's stomach, minimizing the contamination of plants themselves.

Another development being persued is a network of nanosensors and dispensers used throughout a farm field. The sensors recognize when a plant needs nutrients or water, before there is any sign that the plant is deficient. The dispensers then release fertilizer, nutrients, or water as needed, optimizing the growth of each plant in the field one by one.

FUEL CELLS

Nanotechnology is being used to reduce the cost of catalysts used in fuel cells to produce hydrogen ions from fuel such as methanol and to improve the efficiency of membranes used in fuel cells to separate hydrogen ions from other gases such as oxygen.

Fuel Cells and Nanotechnology

Nanotechnology Improve Fuel Cells

Catalysts are used with fuels such as hydrogen or methanol to produce hydrogen ions. Platinum, which is very expensive, is the catalyst typically used in this process. Companies are using nanoparticles of platinum to reduce the amount of platinum needed, or using nanoparticles of other materials to replace platinum entirely and thereby lower costs. Fuel cells contain membranes that allow hydrogen ions to pass through the cell but do not allow other atoms or ions, such as oxygen, to pass through.

Companies are using nanotechnology to create more efficient membranes; this will allow them to build lighter weight and longer lasting fuel cells. Small fuel cells are being developed that can be used to replace batteries in handheld devices such as PDAs or laptop computers.

Most companies working on this type of fuel cell are using methanol as a fuel and are calling them DMFC's, which stands for direct methanol fuel cell.

DMFC's are designed to last longer than conventional batteries. In addition, rather than plugging your device into an electrical outlet and waiting for the battery to recharge, with a DMFC you simply insert a new cartridge of methanol into the device and you're ready to go.

Fuel cells that can replace batteries in electric cars are also under development. Hydrogen is the fuel most researchers propose for use in fuel cell powered cars.

In addition to the improvements to catalysts and membranes discussed above, it is necessary to develop a lightweight and safe hydrogen fuel tank to hold the fuel and build a network of refueling stations.

To build these tanks, researchers are trying to develop lightweight nanomaterials that will absorb the hydrogen and only release it when needed.

The Department of Energy is estimating that widespread usage of hydrogen powered cars will not occur until approximately 2020.

SOLAR CELLS

Companies have developed nanotech solar cells that can be manufactured at significantly lower cost than conventional solar cells.

Solar Cells and Nanotechnology

How can Nanotechnology Improve Solar Cells

Using nanoparticles in the manufacture of solar cells has the following benefits: Reduced manufacturing costs as a result of using a low temperature process similar to printing instead of the high temperature vacuum deposition process typically used to produce conventional cells made with crystalline semiconductor material.

Reduced installation costs achieved by producing flexible rolls instead of rigid crystalline panels. Cells made from semiconductor thin films will also have this characteristic.

Currently available nanotechnology solar cells are not as efficient as traditional ones, however their lower cost offsets this. In the long term nanotechnology versions should both be lower cost and, using quantum dots, should be able to reach higher efficiency levels than conventional ones.

Solar Cells: Nanotechnology Applications under Development

Titanium dioxide nanotubes filled with a polymer to form low cost solar cells Combining carbon nanotubes, bucky-balls and polymers to produce inexpensive solar cells that can be formed by simply painting a surface.

Nanoparticles in plastic film to form solar cells that can be incorporated into cases for devices such as mobile phones and laptop computers. Semiconductor nanoparticles applied in a low temperature printing process that results in low cost solar cells.

Organic molecules to lower costs: Using light absorbing nanowires grown on a flexible film is another method being developed to produce low cost flexible solar panels.

BATTERIES

Companies are currently developing batteries using nanomaterials. One such battery will be a good as new after sitting on the shelf for decades. Another battery can be recharged significantly faster than conventional batteries.

Nanotechnology Battery (Nano Battery)

Using nanotechnology in the manufacture of batteries offers the following benefits: Increasing the available power from a battery and decreasing the time required to recharge a battery. These benefits are achieved by coating the surface of an electrode with nanoparticles.

This increases the surface area of the electrode thereby allowing more current to flow between the electrode and the chemicals inside the battery. This technique could increase the efficiency of hybrid vehicles by significantly reducing the weight of the batteries needed to provide adequate power. Increasing the shelf life of a battery by using nanomaterials to separate liquids in the battery from the solid electrodes when there is no draw on the battery. This separation prevents the low level discharge that occurs in a conventional battery, which increases the shelf life of the battery dramatically. Reducing the possibility of batteries catching fire by providing less flammable electrode material.

Batteries: Nanotechnology Applications under Development

Battery anodes with silicon nanowires that can increase the capacity of Li-ion batteries. Battery small enough to be implanted in the eye and power artificial retina Long shelf life battery uses "nanograss" to separate liquid electrolytes from the solid electrode until power is needed. Lithium ion batteries with nanoparticle (Nano-phosphate™) electrodes that meet the safety requirements for electric cars while improving the perfor-mance. Lithium ion batteries with electrodes made from nano-structured lithium titanate that significantly improves the charge/discharge capability at sub freezing temperatures as well as increasing the upper temperature limit at which the battery remains safe from thermal runaway.

Ultracapacitors using nanotubes may do even better than batteries in hybrid cars. Scientists are using new processes to enhance the storage capacity of the devices. They are banking upon millions of identical nanostructures having peculiar shapes that will facilitate energy transport with the help of electrons. Electrons will move to and fro and store energy at a very large surface area. We all are familiar with the fact that materials behave according to physical laws of nature. The Maryland researchers are using this fact to their advantage. They are utilizing unusual combinations of these behaviors to produce millions and in the end billions of tiny, virtually indistinguishable nanostructures.

Company	Product	Advantages
A123Systems	Lithium-ion battery with the cathode made from nano-phosphate, literature is unclear as to whether this is nanoparticles of phosphate on a substrate or a nano-porous phosphate structure	Higher power, quicker recharge, less combustible than standard lithium-ion batteries
NanoEner Technologies	Electrodes composed of nanoparticles on a substrate for use in batteries. Partner company Enerdel is developing Li Ion battery packs for use in electric and hybrid vehicles	Faster charge and discharge rate than conventional electrodes
Mphase Technologies	Battery with chemicals isolated from electrode by "nanograss" when the battery is not in use	Very long shelf life
Altairnano	Lithium-ion battery with the anode composed of lithium titanate spindel nanoparticles	Higher power, quicker recharge, less combustible than standard lithium-ion batteries
Naoexa	Lithium-ion battery using nanocomposite electrodes using technology developed at Argonne National Laboratory	Higher power, less combustible than standard lithium-ion batteries
EcoloCap Solutions	Lead acid batteries using nanotube coated electrodes	Increased energy density at a lower cost that Li-ion batteries
Zpower	Silver-zinc battery using nanoparticles in the silver cathode	Higher power density, low combustibility
Nexeon	Structured silicon anodes for use in lithium-ion batteries	Higher power density, low combustibility
NanoAmor	Nanotube based additive for use in lithium-ion electrodes	

FUELS

Nanotechnology can address the shortage of fossil fuels such as diesel and gasoline by making the production of fuels from low grade raw materials economical, increasing the mileage of engines, and making the production of fuels from normal raw materials more efficient.

Fuel and Nanotechnology

How can nanotechnology improve fuel availability?

Nanotechnology can address the shortage of fossil fuels such as diesel and gasoline by: Making the production of fuels from low grade raw materials economical Increasing the mileage of engines Making the production of fuels from normal raw materials more efficient Nanotechnology can do all this by increasing the effectiveness of catalysts.

Catalysts can reduce the temperature required to convert raw materials into fuel or increase the Percentage of fuel burned at a given temperature. Catalysts made from nanoparticles have a greater surface area to interact with the reacting chemicals than catalysts made from larger particles.

The larger surface area allows more chemicals to interact with the catalyst simultaneously, which makes the catalyst more effective. This increased effectiveness can make a process such as the production of diesel fuel from coal more economical, and enable the production of fuel from currently unusable raw materials such as low grade crude oil. Nanotechnology, in the form of genetic engineering, can also improve the performance of enzymes used in the conversion of cellulose into ethanol. Currently ethanol added to gasoline in the United States is made from corn, which is driving up the price of corn. The plan is to use engineered enzymes to break down cellulose into sugar, is fermented to turn the sugar into ethanol. This will allow material that often goes to waste, such as wood chips and grass to be turned into ethanol.

BETTER AIR QUALITY

Nanotechnology can improve the performance of catalysts

used to transform vapours escaping from cars or industrial plants into harmless gasses.

That's because catalysts made from nanoparticles have a greater surface area to interact with the reacting chemicals than catalysts made from larger particles. The larger surface area allows more chemicals to interact with the catalyst simultaneously, which makes the catalyst more effective.

How can Nanotechnology Reduce Air Pollution

There are two major ways in which nanotechnology is being used to reduce air pollution: catalysts, which are currently in use and constantly being improved upon; and nano-structured membranes, which are under development.

Catalysts can be used to enable a chemical reaction (which changes one type of molecule to another) at lower temperatures or make the reaction more effective. Nanotechnology can improve the performance and cost of catalysts used to transform vapours escaping from cars or industrial plants into harmless gasses. That's because catalysts made from nanoparticles have a greater surface area to interact with the reacting chemicals than catalysts made from larger particles. The larger surface area allows more chemicals to interact with the catalyst simultaneously, which makes the catalyst more effective. Nanostructured membranes, on the other hand, are being developed to separate carbon dioxide from industrial plant exhaust streams. The plan is to create a method that can be implemented in any power plant without expensive retrofitting.

Cleaner Water

Nanotechnology is being used to develop solutions to three very different problems in water quality. One challenge is the removal of industrial wastes, such as a cleaning solvent called TCE, from groundwater. Nanoparticles can be used to convert the contaminating chemical through a chemical reaction to make it harmless. Studies have shown that this method can be used successfully to reach contaminates dispersed in underground ponds and at much lower cost than methods which require pumping the water out of the ground for treatment.

Reduces Water Pollution

Nanotechnology is being used to develop solutions to three very different problems in water quality. One challenge is the removal of industrial water pollution, such as a cleaning solvent called TCE, from ground water. Nanoparticles can be used to convert the contaminating chemical through a chemical reaction to make it harmless. Studies have shown that this method can be used successfully to reach contaminates dispersed in underground ponds and at much lower cost than methods which require pumping the water out of the ground for treatment.

The challenge is the removal of salt or metals from water. A deionization method using electrodes composed of nano-sized fibers shows promise for reducing the cost and energy requirements of turning salt water into drinking water.

The third problem concerns the fact that standard filters do not work on virus cells. A filter only a few nanometers in diameter is currently being developed that should be capable of removing virus cells from water.

CHEMICAL SENSORS

Nanotechnology can enable sensors to detect very small amounts of chemical vapours. Various types of detecting elements, such as carbon nanotubes, zinc oxide nanowires or palladium nanoparticles can be used in nanotechnology-based sensors.

Because of the small size of nanotubes, nanowires, or nanoparticles, a few gas molecules are sufficient to change the electrical properties of the sensing elements. This allows the detection of a very low concentration of chemical vapours.

Improving Chemical Vapour Sensors

Nanotechnology can enable sensors to detect very small amounts of chemical vapours. Various types of detecting elements, such as carbon nanotubes, zinc oxide nanowires or palladium nanoparticles can be used in nanotechnology-based sensors.

These detecting elements change their electrical characteristics, such as resistance or capacitance, when they absorb a gas molecule.

Because of the small size of nanotubes, nanowires, or nanoparticles, a few gas molecules are sufficient to change the electrical properties of the sensing elements.

This allows the detection of a very low concentration of chemical vapours. The goal is to have small, inexpensive sensors that can sniff out chemicals just as dogs are used in airports to smell the vapours given off by explosives or drugs.

The capability of producing small, inexpensive sensors that can quickly identify a chemical vapour provides a kind of nano-bloodhound that doesn't need sleep or exercise which can be useful in a number of ways.

An obvious application is to mount these sensors throughout an airport, or any facility with security concerns, to check for vapours given off by explosive devices.

These sensors can also be useful in industrial plants that use chemicals in manufacturing to detect the release of chemical vapours.

When hydrogen fuel cells come into use, in cars or other applications, a sensor that detects escaped hydrogen could be very useful in warning of a leak.

This technology should also make possible inexpensive networks of air quality monitoring stations to improve the tracking of air pollution sources.

Nanotechnology Applications Under Development

Hydrogen sensor using a layer of closely spaced palladium nanoparticles that are formed by a beading action like water on a windshield.

When hydrogen is absorbed the palladium nanoparticles swell, causing shorts between nanoparticles which lowers the resistance of the palladium layer. Sensors using zinc oxide nano-wire detection elements capable of detecting a range of chemical vapours.

SPORTING GOODS

If you're a tennis or golf fan, you'll be glad to hear that even sporting goods has wandered into the nano realm. Current nanotechnology applications in the sports arena include increasing the strength of tennis racquets, filling any imperfections in club shaft materials and reducing the rate at which air leaks from tennis balls.

SPORTING GOODS WITH NANOTECHNOLOGY

Nanotechnology Improve Sporting Goods

If you're a tennis or golf fan, you'll be glad to hear that even sporting goods has wandered into the nano realm. *Current nanotechnology applications in the sports arena include*:

- Increasing the strength of tennis racquets by adding nanotubes to the frames which increases control and power when you hit the ball.
- Filling any imperfections in club shaft materials with nanoparticles; this improves the uniformity of the material that makes up the shaft and thereby improving your swing.
- Reducing the rate at which air leaks from tennis balls so they keep their bounce longer.

FABRIC

Making composite fabric with nano-sized particles or fibers allows improvement of fabric properties without a significant increase in weight, thickness, or stiffness as might have been the case with previously-used techniques.

Nanotechnology Improve Fabric

Making composite fabric with nano-sized particles or fibers allows improvement of fabric properties without a significant increase in weight, thickness, or stiffness as might have been the case with previously-used techniques. For example incorporating nano-whiskers into fabric used to make pants produces a lightweight water and stain repellent material.

Fabric: Current Nanotechnology Applications

- Silver nanoparticles in fabric that kills bacteria making clothing odor- resistant.
- Nanopores providing superior insulation for shoe inserts in cold weather.

Chapter 2

Physics of Solid State

STRUCTURE

SIZE DEPENDENCE OF PROPERTIES

Properties of solids determined by their shapes and sizes are indeed fascinating, which form the basis of Nanoscience and Nanotechnology that has been recognized as key significance in Science, Technology, and Economics in the 21 st Century. Compared to the overwhelming contributions to nanotechnology, fundamental insight into the mechanism behind the nanosolid tunability remains yet its infancy.

The classical theories in terms of continuum mechanics and statistic thermodynamics exhibit limitations. For a given phenomenon, there are often numerous theories discussing from various top-down perspectives.

A theory is yet lacking to systematically unify all the observed tunability of nanosolids in a bottom-up approach in a predictive way. Furthermore, structural miniaturization provides us with a new freedom that not only allows us to tune the properties of a solid, but also challenges us to gain information that is beyond traditional approaches. The BOLS correlation mechanism is established under these circumstances.

The BOLS correlation theory indicates that the bond order loss of atoms at sites surrounding defects or near the edge of a surface causes the remaining bonds of the under-coordinated atom to contract spontaneously associated with bond strength gain, or intra-atomic potential well depression. Consequently, localized densification of charge, mass, and energy occurs in the relaxed skin, which perturbs the Hamiltonian (energy density), atomic cohesive energy (the product of single bond energy and the number of bond of a specific atom), and electroaffinity (separation between the vacuum level and the conduction band edge), and the associated properties of the solid, as can be observed experimentally.

For instances, the entire band structure such as the band-gap widths, core-level shifts, Stokes shift (electron-phonon interaction), and dielectrics depend functionally on the Hamiltonian.

The modified atomic cohesive energy dictates the thermodynamic process of the solid such as self-assembly growth, atomic vibration, phase transition, diffusivity, sinter-ability, and thermal stability.

The enhanced electroaffinity determines the chemical reactivity and toxity of nanostructures. The trapping potential well depression influences the transport dynamics of phonons, photons, and electrons in the grain boundaries and the entire solid.

If the functional dependence of the detectable quantities such as the Young's modulus and dielectric constant on the bond order, bond length, and bond strength and their coordination and temperature dependence are established, we can then readily predict the performance with the number of the under-coordinated atoms of a system.

CRYSTAL STRUCTURE

In any sort of discussion of crystalline materials, it is useful to begin with a discussion of crystallography: the study of the formation, structure, and properties of crystals. A crystal structure is defined as the particular repeating arrangement of atoms (molecules or ions) throughout a crystal.

Structure refers to the internal arrangement of particles and not the external appearance of the crystal. However, these are not entirely independent since the external appearance of a crystal is often related to the internal arrangement. For example, crystals of cubic rock salt (NaCl) are physically cubic in appearance. Only a few of the possible crystal structures are of concern with respect to simple inorganic salts and these will be discussed in detail, however, it is important to understand the nomenclature of crystallography.

Crystallography

Bravais Lattice

The Bravais lattice is the basic building block from which all crystals can be constructed. The concept originated as a topological problem of finding the number of different ways to arrange points in space where each point would have an identical "atmosphere". That is each point would be surrounded by an identical set of points as any other point, so that all points would be indistinguishable from each other.

Mathematician Auguste Bravais discovered that there were 14 different collections of the groups of points, which are known as Bravais lattices. These lattices fall into seven different "crystal systems", as differentiated by the relationship between the angles between sides of the "unit cell" and the distance between points in the unit cell. The unit cell is the smallest group of atoms, ions or molecules that, when repeated at regular intervals in three dimensions, will produce the lattice of a crystal system.

The "lattice parameter" is the length between two points on the corners of a unit cell. Each of the various lattice parameters are designated by the letters *a*, *b*, and *c*. If two sides are equal, such as in a tetragonal lattice, then the lengths of the two lattice parameters are designated*a* and *c*, with *b* omitted. The angles are designated by the Greek letters α, β, and *γ*, such that an angle with a specific Greek letter is not subtended by the axis with its Roman equivalent. For example, α is the included angle between the *b*and *c* axis.

Table below shows the various crystal systems, while Figure shows the 14 Bravais lattices. It is important to distinguish the characteristics of each of the individual systems. An example of a material that takes on each of the Bravais lattices is shown in next Table.

Table: Geometrical Characteristics of the Seven Crystal Systems.

1. *System*: Cubic
   Axial lengths and angles: $a = b = c$, $\alpha = \beta = \gamma = 90°$
   Unit cell geometry:

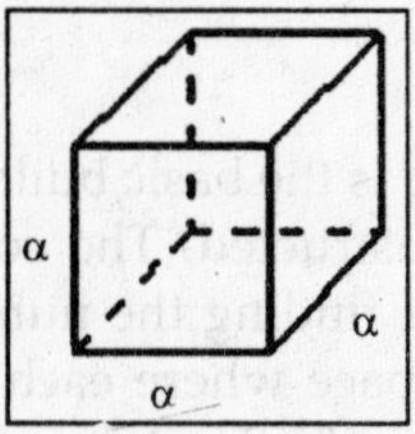

2. *System*: Cubic
   Axial lenhs and angles: $a = b = c$, $\alpha = \beta = \gamma$ 90°

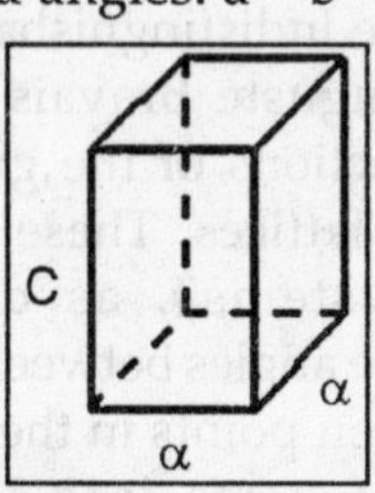

3. *System*: Tetragonal
   Axial lenhs and angles: $a = b \neq c$, $\alpha = \beta = \gamma = 90°$

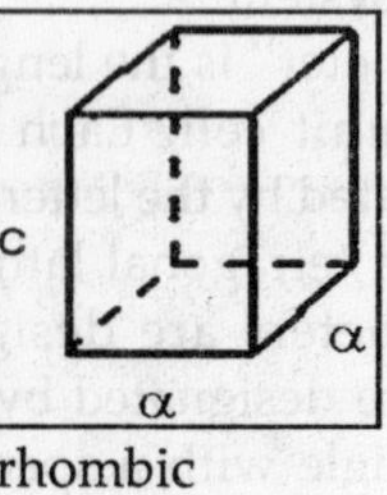

4. *System*: Orthorhombic
   Axial lenhs and angles: $a \neq b \neq c$, $\alpha = \beta = \gamma$ 90°

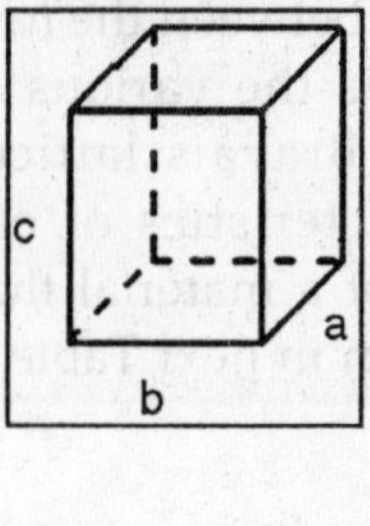

5. *System*: Rhombohedral
   Axial lenhs and angles: $a = b = c, \alpha = \beta = \gamma \neq 90°$

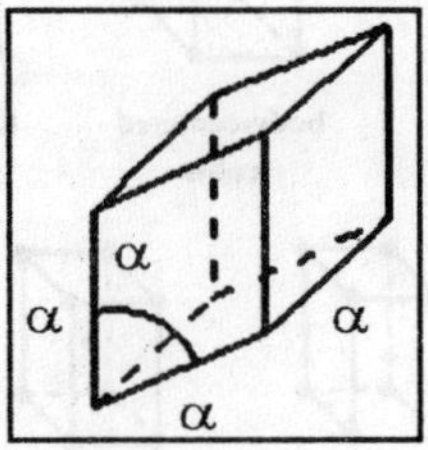

6. *System*: Hexagonal
   Axial lenhs and angles: $a = b \neq c, \alpha = \beta = 90°, \gamma = 120°$

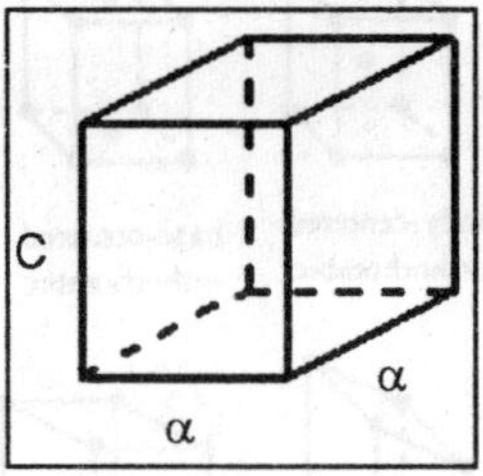

7. *System*: Monoclinic
   Axial lenhs and angles: $a \neq b \neq c, \alpha = \gamma = 90°, \beta \neq 90°$

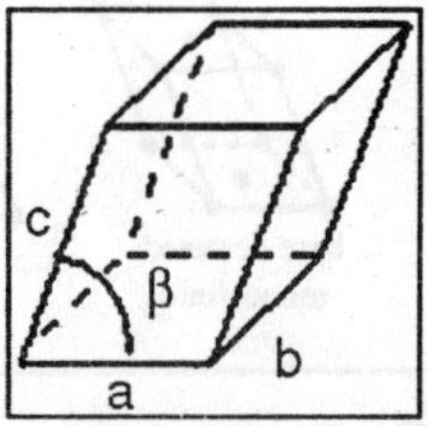

8. *Axial lenhs and angles*: $a \neq b \neq c, \alpha \neq \beta \neq \gamma$

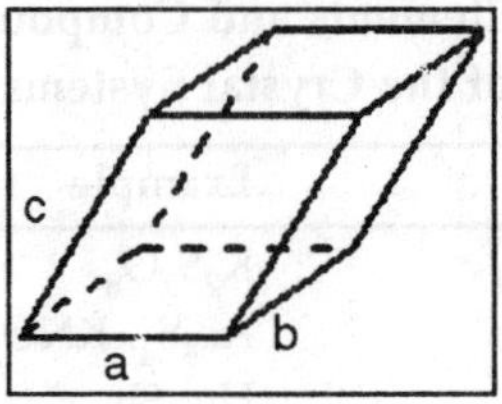

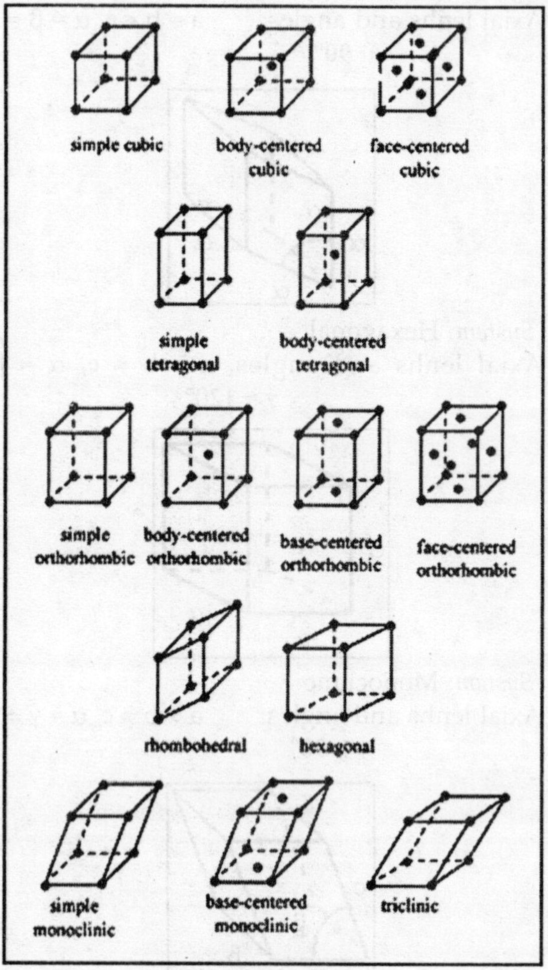

Fig. Bravais Lattices.

Table. Examples of Elements and Compounds that Adopt Each of the Crystal Systems.

Crystal system	Example
Triclinic	$K_2S_2O_8$
Monoclinic	As_4S_4, KNO_2
Rhombohedral	Hg, Sb

Hexagonal	Zn, Co, NiAs
Orthorhombic	Ga, Fe_3C
Tetragonal	In, TiO_2
Cubic	Au, Si, NaCl

The cubic lattice is the most symmetrical of the systems. All the angles are equal to 90°, and all the sides are of the same length ($a = b = c$). Only the length of one of the sides (a) is required to describe this system completely.

In addition to simple cubic, the cubic lattice also includes body-Centreed cubic and face-Centreed cubic . Body-Centreed cubic results from the presence of an atom (or ion) in the Centre of a cube, in addition to the atoms (ions) positioned at the vertices of the cube. In a similar manner, a face-Centreed cubic requires, in addition to the atoms (ions) positioned at the vertices of the cube, the presence of atoms (ions) in the Centre of each of the cubes face.

The tetragonal lattice has all of its angles equal to 90°, and has two out of the three sides of equal length ($a = b$). The system also includes body-Centreed tetragonal . In an orthorhombic lattice all of the angles are equal to 90°, while all of its sides are of unequal length.

The system needs only to be described by three lattice parameters. This system also includes body-Centreed orthorhombic, base-Centreed orthorhombic, and face-Centreed orthorhombic (Figure). A base-Centreed lattice has, in addition to the atoms (ions) positioned at the vertices of the orthorhombic lattice, atoms (ions) positioned on just two opposing faces.

The rhombohedral lattice is also known as trigonal, and has no angles equal to 90°, but all sides are of equal length ($a = b = c$), thus requiring only by one lattice parameter, and all three angles are equal ($\alpha = \beta = \gamma$).

A hexagonal crystal structure has two angles equal to 90°, with the other angle (γ) equal to 120°. For this to happen, the two sides surrounding the 120° angle must be equal ($a = b$), while the third side (c) is at 90° to the other sides and can be of any length.

The monoclinic lattice has no sides of equal length, but

two of the angles are equal to 90°, with the other angle (usually defined as β) being something other than 90°. It is a tilted parallelogram prism with rectangular bases. This system also includes base-Centreed monoclinic.

In the triclinic lattice none of the sides of the unit cell are equal, and none of the angles within the unit cell are equal to 90°. The triclinic lattice is chosen such that all the internal angles are either acute or obtuse.

This crystal system has the lowest symmetry and must be described by 3 lattice parameters (*a*, *b*, and *c*) and the 3 angles (α, β, and γ).

Atom Positions, Crystal Directions and Miller Indices

Atom Positions and Crystal Axes

The structure of a crystal is defined with respect to a unit cell. As the entire crystal consists of repeating unit cells, this definition is sufficient to represent the entire crystal. Within the unit cell, the atomic arrangement is expressed using coordinates.

There are two systems of coordinates commonly in use, which can cause some confusion. Both use a corner of the unit cell as their origin.

The first, less-commonly seen system is that of Cartesian or orthogonal coordinates (X, Y, Z). These usually have the units of Angstroms and relate to the distance in each direction between the origin of the cell and the atom.

These coordinates may be manipulated in the same fashion are used with two- or three-dimensional graphs. It is very simple, therefore, to calculate inter-atomic distances and angles given the Cartesian coordinates of the atoms.

Unfortunately, the repeating nature of a crystal cannot be expressed easily using such coordinates. For example, consider a cubic cell of dimension 3.52 Å. Pretend that this cell contains an atom that has the coordinates (1.5, 2.1, 2.4).

That is, the atom is 1.5 Å away from the origin in the x direction (which coincides with the *a* cell axis), 2.1 Å in the y (which coincides with the *b* cell axis) and 2.4 Å in the z (which coincides with the *c* cell axis). There will be an equivalent atom in the next unit cell along the x-direction, which will have the coordinates (1.5 + 3.52, 2.1, 2.4) or (5.02, 2.1, 2.4).

This was a rather simple calculation, as the cell has very high symmetry and so the cell axes, *a, b* and *c*, coincide with the Cartesian axes, X, Y and Z. However, consider lower symmetry cells such as triclinic or monoclinic in which the cell axes are not mutually orthogonal. In such cases, expressing the repeating nature of the crystal is much more difficult to accomplish.

Accordingly, atomic coordinates are usually expressed in terms of fractional coordinates, (x, y, z). This coordinate system is coincident with the cell axes (*a, b, c*) and relates to the position of the atom in terms of the fraction along each axis.

Consider the atom in the cubic cell discussion above. The atom was 1.5 Å in the *a* direction away from the origin. As the *a* axis is 3.52 Å long, the atom is ($^{1.5}/_{3.52}$) or 0.43 of the axis away from the origin.

Similarly, it is ($^{2.1}/_{3.52}$) or 0.60 of the *b* axis and ($^{2.4}/_{3.5}$) or 0.68 of the *c* axis. The fractional coordinates of this atom are, therefore, (0.43, 0.60, 0.68).

The coordinates of the equivalent atom in the next cell over in the *a* direction, however, are easily calculated as this atom is simply 1 unit cell away in *a*.

Thus, all one has to do is add 1 to the x coordinate: (1.43, 0.60, 0.68). Such transformations can be performed regardless of the shape of the unit cell. Fractional coordinates, therefore, are used to retain and manipulate crystal information.

Crystal directions

The designation of the individual vectors within any given crystal lattice is accomplished by the use of whole number multipliers of the lattice parameter of the point at which the vector exits the unit cell.

The vector is indicated by the notation [*hkl*], where *h, k,* and *l* are reciprocals of the point at which the vector exits the unit cell.

The origination of all vectors is assumed defined as [000]. For example, the direction along the *a*-axis according to this scheme would be [100] because this has a component only in the *a*-direction and no component along either the *b* or *c* axial direction.

A vector diagonally along the face defined by the *a* and *b* axis would be, while going from one corner of the unit cell to the opposite corner would be in the direction shows some examples of the various directions in the unit cell.

The crystal direction notation is made up of the lowest combination of integers and represents unit distances rathe: than actual distances.

A direction is identical to a, so is used. Fractions are not used. For example, a vector that intercepts the Centre of the top face of the unit cell has the coordinates x = 1/2, y = 1/2, z = 1. All have to be inversed to convert to the lowest combination of integers (whole numbers); i.e..

Finally, all parallel vectors have the same crystal direction, e.g., the four vertical edges of the cell shown in all have the crystal direction [*hkl*] = [001].

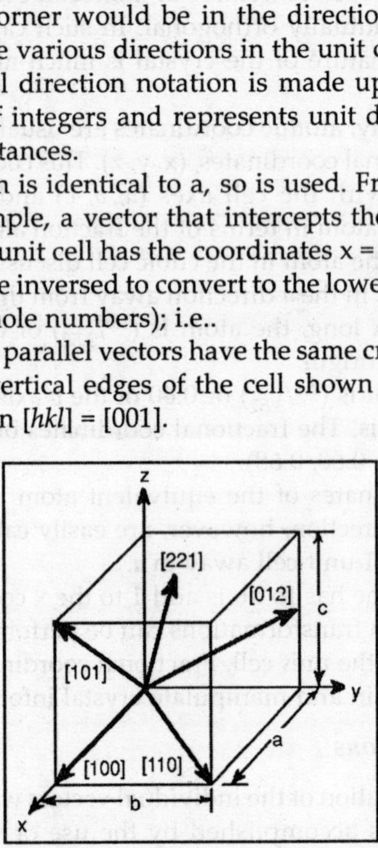

Fig. Some Common Directions in a Cubic Unit Cell.

Crystal directions may be grouped in families. To avoid confusion there exists a convention in the choice of brackets surrounding the three numbers to differentiate a crystal direction from a family of direction.

For a direction, square brackets [*hkl*] are used to indicate an individual direction. Angle brackets <*hkl*> indicate a family of directions. A family of directions includes any directions that are equivalent in length and types of atoms encountered. For example, in a cubic lattice, the directions all belong to the <100> family of planes because they are equivalent.

If the cubic lattice were rotated 90°, the *a*, *b*, and *c* directions would remain indistinguishable, and there would be no way of telling on which crystallographic positions the atoms are situated, so the family of directions is the same.

In a hexagonal crystal, however, this is not the case, so the and would both be <100> directions, but the direction would be distinct. Finally, negative directions are identified with a bar over the negative number instead of a minus sign.

Crystal planes

Planes in a crystal can be specified using a notation called Miller indices. The Miller index is indicated by the notation [*hkl*] where *h*, *k*, and *l* are reciprocals of the plane with the *x*, *y*, and z axes.

To obtain the Miller indices of a given plane requires the following steps:

Step 1. The plane in question is placed on a unit cell.

Step 2. Its intercepts with each of the crystal axes are then found.

Step 3. The reciprocal of the intercepts are taken.

Step 4. These are multiplied by a scalar to insure that is in the simple ratio of whole numbers.

For example, the face of a lattice that does not intersect the y or z axis would be, while a plane along the body diagonal would be the plane.

As with crystal directions, Miller indices directions may be grouped in families. Individual Miller indices are given in parentheses (*hkl*), while braces {*hkl*} are placed around the indices of a family of planes.

For example, (001), (100), and (010) are all in the {100} family of planes, for a cubic lattice. An illustration of this along with the and planes is given in figure.

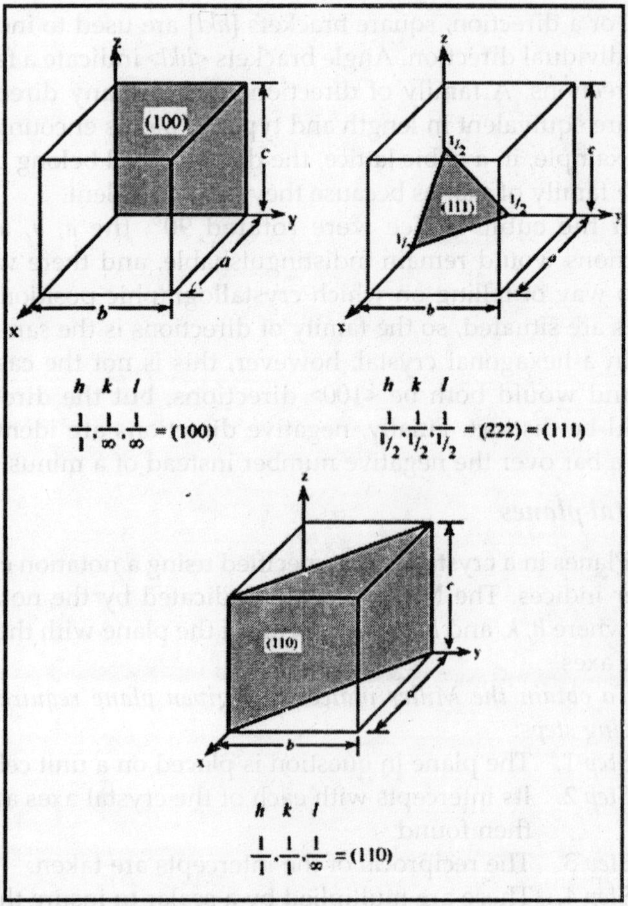

Fig. Examples of Miller Indices Notation for Crystal Planes.

Description of Crystal Structures

Crystal structures may be described in a number of ways. The most common manner is to refer to the size and shape of the unit cell and the positions of the atoms (or ions) within the cell.

However, this information is sometimes insufficient to allow for an understanding of the true structure in three dimensions.

Consideration of several unit cells, the arrangement of the atoms with respect to each other, the number of other atoms they in contact with, and the distances to neighboring atoms, often will provide a better understanding.

A number of methods are available to describe extended solid-state structures. The most applicable with regard to elemental and compound semiconductor, metals and the majority of insulators is the close packing approach.

Many crystal structures can be described using the concept of close packing. This concept requires that the atoms (ions) are arranged so as to have the maximum density. In order to understand close packing in three dimensions, the most efficient way for equal sized spheres to be packed in two dimensions must be considered.

Close Packed Structures: Hexagonal close Packing and Cubic close Packing

The most efficient way for equal sized spheres to be packed in two dimensions is shown in Figure , in which it can be seen that each sphere (the dark gray shaded sphere) is surrounded by, and is in contact with, six other spheres (the light gray spheres in Figure).

It should be noted that contact with six other spheres the maximum possible is the spheres are the same size, although lower density packing is possible. Close packed layers are formed by repetition to an infinite sheet. Within these close packed layers, three close packed rows are present, shown by the dashed lines in Figure.

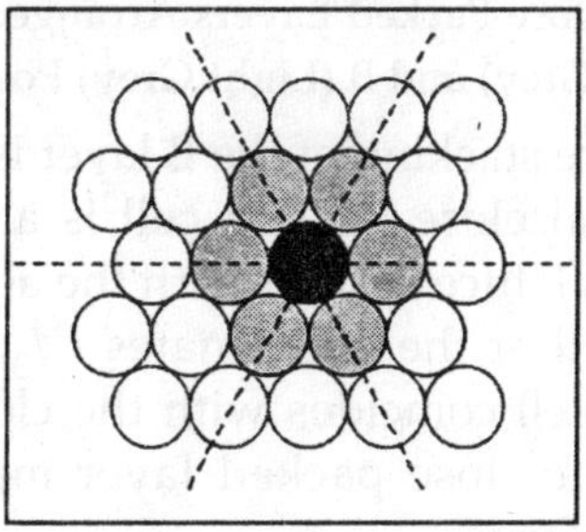

Fig. Close Packed Layer of Equal Sized Spheres.

The close packed rows (directions) are shown by the dashed lines. The most efficient way for equal sized spheres to be packed in three dimensions is to stack close packed layers on top of each other to give a close packed structure. There are two simple ways in which this can be done, resulting in either a hexagonal or cubic close packed structures.

Hexagonal Close Packed

If two close packed layers A and B are placed in contact with each other so as to maximize the density, then the spheres of layer B will rest in the hollow (vacancy) between three of the spheres in layer A.

This is demonstrated in figure. Atoms in the second layer, B (shaded light gray), may occupy one of two possible positions but not both together or a mixture of each.

If a third layer is placed on top of layer B such that it exactly covers layer A, subsequent placement of layers will result in the following sequence...ABABAB.... This is known as hexagonal close packing or *hcp*.

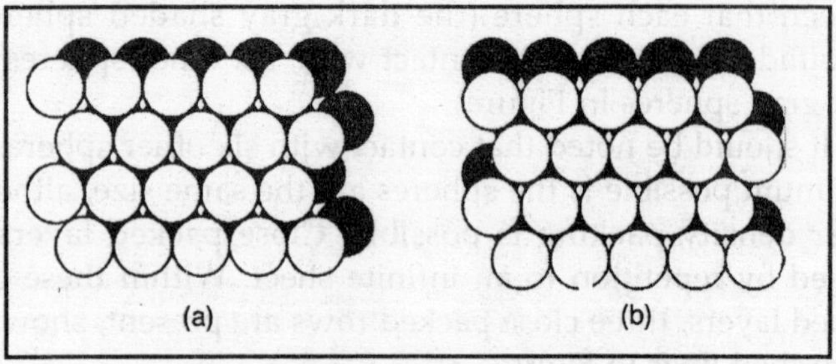

Fig. Close Packed Layers Arranged in A (Dark Grey) and B (Light Grey) Positions.

The alternative stacking of the B layer is shown in (a) and (b). The hexagonal close packed cell is a derivative of the hexagonal Bravais lattice system with the addition of an atom inside the unit cell at the coordinates ($^1/_3$,$^2/_3$,$^1/_2$). The basal plane of the unit cell coincides with the close packed layers. In other words the close packed layer makes-up the {001} family of crystal planes.

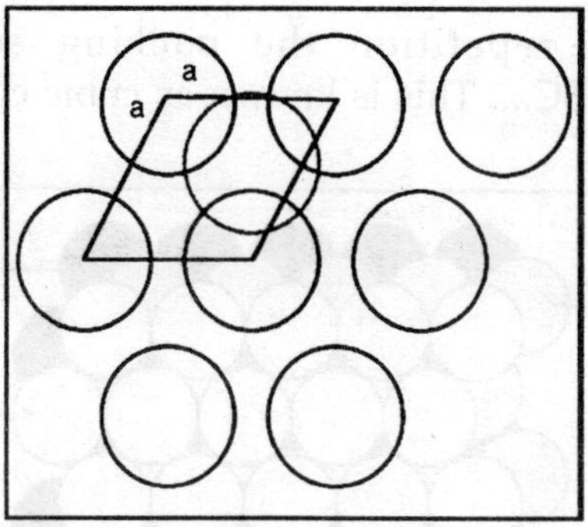

Fig. Basal Plane of Hcp Unit Cell on the Close Packed Layers.

The "packing fraction" in a hexagonal close packed cell is 74.05%; that is 74.05% of the total volume is occupied. The packing fraction or density is derived by assuming that each atom is a hard sphere in contact with its nearest neighbors.

Determination of the packing fraction is accomplished by calculating the number of whole spheres per unit cell (2 in hcp), the volume occupied by these spheres, and a comparison with the total volume of a unit cell.

The number gives an idea of how "open" or filled a structure is. By comparison, the packing fraction for body-Centreed cubic is 68% and for diamond cubic (an important semiconductor structure to be described later) is it 34%.

Cubic Close Packed: Face-Centreed Cubic

The unit cell of cubic close packed structure is actually that of a face-Centreed cubic (*fcc*) Bravais lattice. In the *fcc*lattice the close packed layers constitute the {111} planes.

As with the *hcp* lattice packing fraction in a cubic close packed (*fcc*) cell is 74.05%. Since face Centreed cubic or *fcc* is more commonly used in preference to cubic close packed (*ccp*) in describing the structures, the former will be used throughout this text.

In a similar manner to the generation of the hexagonal close packed structure, two close packed layers are stacked however, the third layer (C) is placed such that it does not exactly cover layer A, while sitting in a set of troughs in layer

B then upon repetition the packing sequence will be...ABCABCABC.... This is known as cubic close packing or *ccp*.

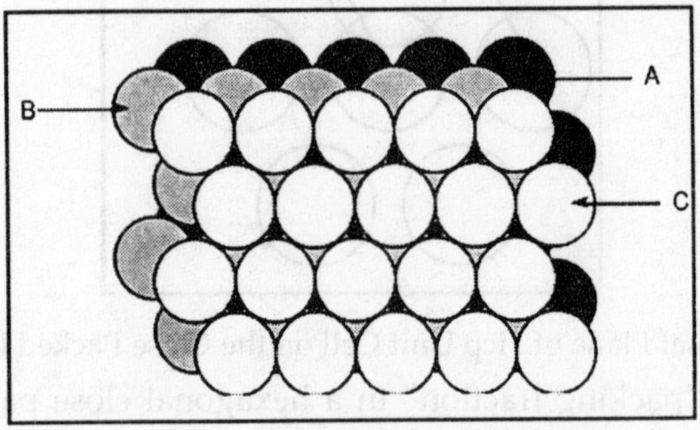

Fig. Close Packed Layers in a Cubic Close Packed Arrangement:

Coordination Number

The coordination number of an atom or ion within an extended structure is defined as the number of nearest neighbour atoms (ions of opposite charge) that are in contact with it. A slightly different definition is often used for atoms within individual molecules: the number of donor atoms associated with the central atom or ion. However, this distinction is rather artificial, and both can be employed.

The coordination numbers for metal atoms in a molecule or complex are commonly 4, 5, and 6, but all values from 2 to 9 are known and a few examples of higher coordination numbers have been reported. In contrast, common coordination numbers in the solid state are 3, 4, 6, 8, and 12. For example, the atom in the Centre of body-Centreed cubic lattice has a coordination number of 8, because it touches the eight atoms at the corners of the unit cell, while an atom in a simple cubic structure would have a coordination number of 6. In both *fcc* and *hcp* lattices each of the atoms have a coordination number of 12.

Octahedral and Tetrahedral Vacancies

As was mentioned above, the packing fraction in both *fcc*

and *hcp* cells is 74.05%, leaving 25.95% of the volume unfilled. The unfilled lattice sites (interstices) between the atoms in a cell are called interstitial sites or vacancies. The shape and relative size of these sites is important in controlling the position of additional atoms.

In both *fcc* and *hcp* cells most of the space within these atoms lies within two different sites known as octahedral sites and tetrahedral sites. The difference between the two lies in their "coordination number", or the number of atoms surrounding each site.

Tetrahedral sites (vacancies) are surrounded by four atoms arranged at the corners of a tetrahedron. Similarly, octahedral sites are surrounded by six atoms which make-up the apices of an octahedron. For a given close packed lattice an octahedral vacancy will be larger than a tetrahedral vacancy.

Within a face Centreed cubic lattice, the eight tetrahedral sites are positioned within the cell, at the general fractional coordinate of $({}^{n}/_{4},{}^{n}/_{4},{}^{n}/_{4})$ where n = 1 or 3, e.g., $({}^{1}/_{4},{}^{1}/_{4},{}^{1}/_{4})$, $({}^{1}/_{4},{}^{1}/_{4},{}^{3}/_{4})$, etc. The octahedral sites are located at the Centre of the unit cell $({}^{1}/_{2},{}^{1}/_{2},{}^{1}/_{2})$, as well as at each of the edges of the cell, e.g., $({}^{1}/_{2},0,0)$. In the hexagonal close packed system, the tetrahedral sites are at $(0,0,{}^{3}/_{8})$ and $({}^{1}/_{3},{}^{2}/_{3},{}^{7}/_{8})$, and the octahedral sites are at $({}^{1}/_{3},{}^{1}/_{3},{}^{1}/_{4})$ and all symmetry equivalent positions.

Important Structure Types

The majority of crystalline materials do not have a structure that fits into the one atom per site simple Bravais lattice. A number of other important crystal structures are found, however, only a few of these crystal structures are those of which occur for the elemental and compound semiconductors and the majority of these are derived from *fcc* or *hcp* lattices.

Each structural type is generally defined by an archetype, a material (often a naturally occurring mineral) which has the structure in question and to which all the similar materials are related. With regard to commonly used elemental and compound semiconductors the important structures are diamond, zinc blende, Wurtzite, and to a lesser extent chalcopyrite.

However, rock salt, β-tin, cinnabar and cesium chloride are observed as high pressure or high temperature phases and are therefore also discussed.

The following provides a summary of these structures. Details of the full range of solid-state structures are given elsewhere.

Diamond Cubic

The diamond cubic structure consists of two interpenetrating face-Centreed cubic lattices, with one offset $^1/_4$ of a cube along the cube diagonal. It may also be described as face Centreed cubic lattice in which half of the tetrahedral sites are filled while all the octahedral sites remain vacant.

The diamond cubic unit cell is shown in Figure. Each of the atoms (e.g., C) is four coordinate, and the shortest interatomic distance (C-C) may be determined from the unit cell parameter (*a*).

$$C-C = a\frac{\sqrt{3}}{4} \approx 0.422a$$

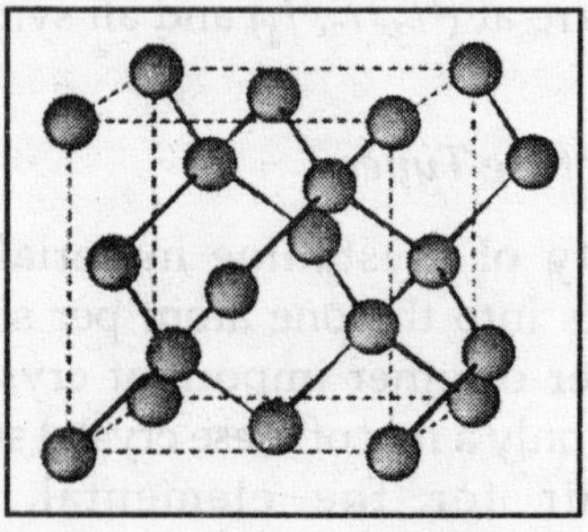

Fig. Unit Cell Structure of a Diamond Cubic Lattice.

Zinc Blende

This is a binary phase (ME) and is named after its archetype, a common mineral form of zinc sulfide (ZnS). As with the diamond lattice, zinc blende consists of the two interpenetrating *fcc* lattices. However, in zinc blende one lattice consists of one of the types of atoms (Zn in ZnS), and the other lattice is of the second type of atom (S in ZnS).

It may also be described as face Centreed cubic lattice of S atoms in which half of the tetrahedral sites are filled with Zn atoms.

All the atoms in a zinc blende structure are 4-coordinate. The zinc blende unit cell is shown in Figure. A number of inter-atomic distances may be calculated for any material with a zinc blende unit cell using the lattice parameter (*a*).

$$\mathrm{Zn-S} = a\frac{\sqrt{3}}{4} \approx 0.422a$$

$$\mathrm{Zn-Zn} = \mathrm{S-S} = \frac{a}{\sqrt{2}} \approx 0.707a$$

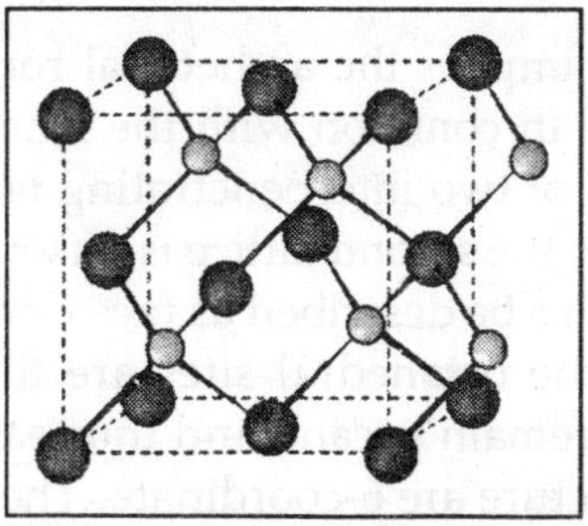

Fig. Unit Cell Structure of a Zinc Blende (ZnS) Lattice.

Zinc atoms are shown in green (small), sulfur atoms shown in red (large), and the dashed lines show the unit cell.

Chalcopyrite

The mineral chalcopyrite $CuFeS_2$ is the archetype of this structure. The structure is tetragonal ($a = b \neq c$, $\alpha = \beta = \gamma = 90°$, and is essentially a superlattice on that of zinc blende.

Thus, is easiest to imagine that the chalcopyrite lattice is made-up of a lattice of sulfur atoms in which the tetrahedral sites are filled in layers...FeCuCuFe..., etc.In such an idealized structure $c = 2a$, however, this is not true of all materials with chalcopyrite structures.

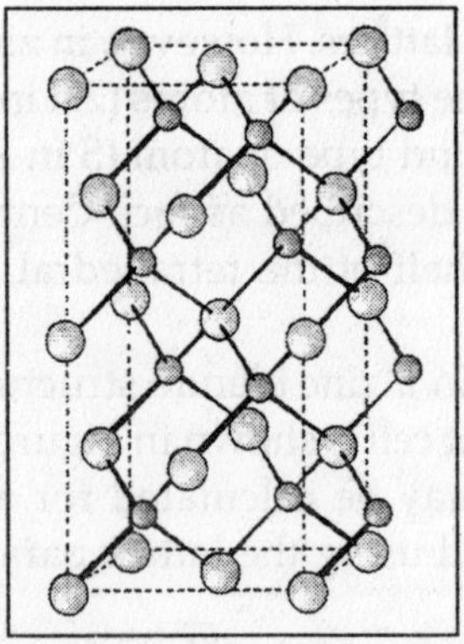

Fig. Unit Cell Structure of a Chalcopyrite Lattice.

Copper atoms are shown in blue, iron atoms are shown in green and sulfur atoms are shown in yellow.

Rock Salt

As its name implies the archetypal rock salt structure is NaCl (table salt). In common with the zinc blende structure, rock salt consists of two interpenetrating face-Centreed cubic lattices. However, the second lattice is offset 1/2*a*along the unit cell axis. It may also be described as face Centreed cubic lattice in which all of the octahedral sites are filled, while all the tetrahedral sites remain vacant, and thus each of the atoms in the rock salt structure are 6-coordinate. The rock salt unit cell is shown in Figure below. A number of inter-atomic distances may be calculated for any material with a rock salt structure using the lattice parameter (*a*).

$$\mathrm{Na-CI} = \frac{a}{2} \approx 0.5a$$

$$\mathrm{Na-Na} = \mathrm{CI-CI} = \frac{a}{\sqrt{2}} \approx 0.707a$$

Sodium ions are shown in purple (small spheres) and chloride ions are shown in red (large spheres).

Cinnabar: Cinnabar, named after the archetype mercury sulfide, HgS, is a distorted rock salt structure in which the resulting cell is rhombohedral (trigonal) with each atom having a coordination number of six.

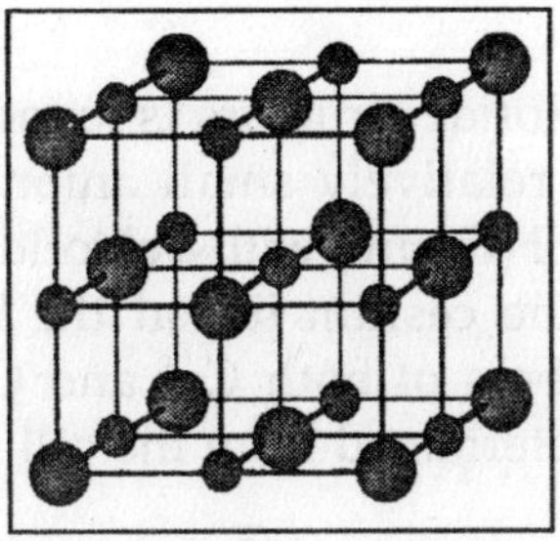

Fig. Unit Cell Structure of a Rock Salt Lattice.

Wurtzite

This is a hexagonal form of the zinc sulfide. It is identical in the number of and types of atoms, but it is built from two interpenetrating *hcp* lattices as opposed to the *fcc* lattices in zinc blende. As with zinc blende all the atoms in a wurtzite structure are 4-coordinate. The wurtzite unit cell is shown in Figure. A number of inter atomic distances may be calculated for any material with a wurtzite cell using the lattice parameter (*a*).

$$\mathrm{Zn-S} = a\sqrt{3}/8 = 0.612a = 3\frac{c}{8} = 0.375c$$

$$\mathrm{Zn–Zn} = \mathrm{S–S} = a = 1.632\ c$$

However, it should be noted that these formulae do not necessarily apply when the ratio *a/c* is different from the ideal value of 1.632.

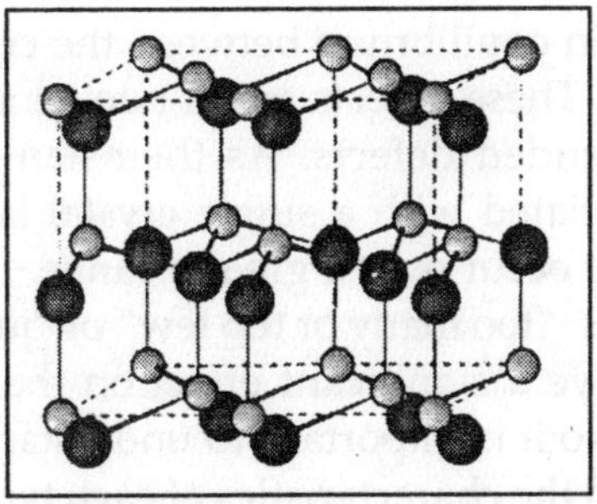

Fig. Unit Cell Structure of a Wurtzite Lattice.

Zinc atoms are shown in green (small spheres), sulfur atoms shown in red (large spheres), and the dashed lines show the unit cell.

Cesium Chloride

The cesium chloride structure is found in materials with large cations and relatively small anions. It has a simple (primitive) cubic cell (Figure) with a chloride ion at the corners of the cube and the cesium ion at the body Centre. The coordination numbers of both Cs^+ and Cl^-, with the inner atomic distances determined from the cell lattice constant (*a*).

$$Cs - CI = \frac{a\sqrt{3}}{2} \approx 0.866a$$

$$Cs - Cs = CI - CI = a$$

β-Tin

The room temperature allotrope of tin is β-tin or white tin. It has a tetragonal structure, in which each tin atom has four nearest neighbors (Sn-Sn = 3.016 Å) arranged in a very flattened tetrahedron, and two next nearest neighbors (Sn-Sn = 3.175 Å). The overall structure of β-tin consists of fused hexagons, each being linked to its neighbour via a four-membered Sn_4 ring.

Defects in crystalline solids

Up to this point we have only been concerned with ideal structures for crystalline solids in which each atom occupies a designated point in the crystal lattice. Unfortunately, defects ordinarily exist in equilibrium between the crystal lattice and its environment. These defects are of two general types: point defects and extended defects. As their names imply, point defects are associated with a single crystal lattice site, while extended defects occur over a greater range.

Point defects: "too many or too few" or "just plain wrong" Point defects have a significant effect on the properties of a semiconductor, so it is important to understand the classes of point defects and the characteristics of each type. Figure below summarizes various classes of native point defects, however, they may be divided into two general classes; defects with the wrong number of atoms (deficiency or surplus) and defects where the identity of the atoms is incorrect.

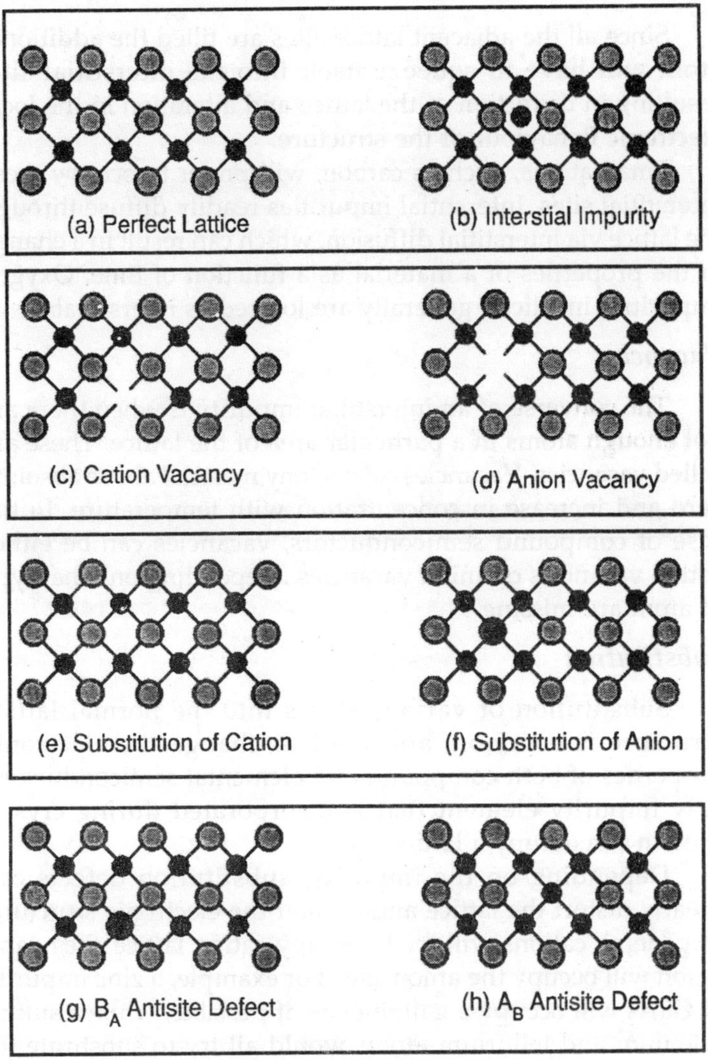

Fig. Point Defects in a Crystal Lattice.

Interstitial Impurity

An interstitial impurity occurs when an extra atom is positioned in a lattice site that should be vacant in an ideal structure .

Since all the adjacent lattice sites are filled the additional atom will have to squeeze itself into the interstitial site, resulting in distortion of the lattice and alteration in the local electronic Behaviour of the structure.

Small atoms, such as carbon, will prefer to occupy these interstitial sites. Interstitial impurities readily diffuse through the lattice via interstitial diffusion, which can result in a change of the properties of a material as a function of time. Oxygen impurities in silicon generally are located as interstitials.

Vacancies

The converse of an interstitial impurity is when there are not enough atoms in a particular area of the lattice. These are called vacancies. Vacancies exist in any material above absolute zero and increase in concentration with temperature. In the case of compound semiconductors, vacancies can be either cation vacancies or anion vacancies , depending on what type of atom are"missing".

Substitution

Substitution of various atoms into the normal lattice structure is common, and used to change the electronic properties of both compound and elemental semiconductors. Any impurity element that is incorporated during crystal growth can occupy a lattice site.

Depending on the impurity, substitution defects can greatly distort the lattice and/or alter the electronic structure. In general, cations will try to occupy cation lattice sites , and anion will occupy the anion site . For example, a zinc impurity in GaAs will occupy a gallium site, if possible, while a sulfur, selenium and tellurium atoms would all try to substitute for an arsenic. Some impurities will occupy either site indiscriminately, e.g., Si and Sn occupy both Ga and As sites in GaAs.

Antisite Defects

Antisite defects are a particular form of substitution defect, and are unique to compound semiconductors. An antisite defect occurs when a cation is misplaced on an anion lattice site or vice versa.

Dependant on the arrange ment these are designated as either AB antisite defects or BA antisite defects. For example, if an arsenic atom is on a gallium lattice site the defect would be an AsGa defect.

Antisite defects involve fitting into a lattice site atoms of a different size than the rest of the lattice, and therefore this often results in a localized distortion of the lattice.

In addition, cations and anions will have a different number of electrons in their valence shells, so this substitution will alter the local electron concentration and the electronic properties of this area of the semiconductor.

Extended Defects: Dislocations in a Crystal Lattice

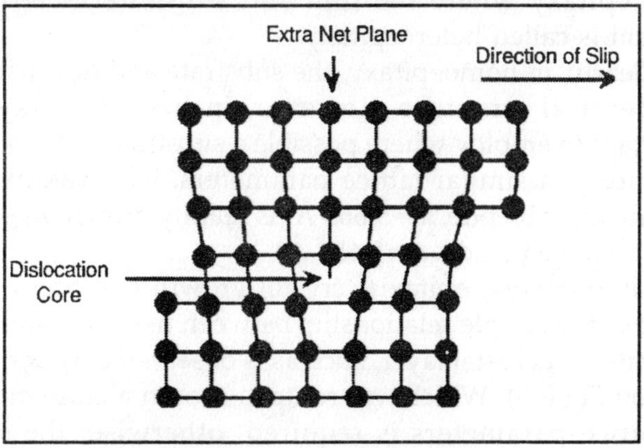

Fig. Dislocation in a Crystal Lattice.

Extended defects may be created either during crystal growth or as a consequence of stress in the crystal lattice. The plastic deformation of crystalline solids does not occur such that all bonds along a plane are broken and reformed simultaneously. Instead, the deformation occurs through a dislocation in the crystal lattice. Figure below shows a schematic representation of a dislocation in a crystal lattice. Two features of this type of dislocation are the presence of an extra crystal plane, and a large void at the dislocation core.

Impurities tend to segregate to the dislocation core in order to relieve strain from their presence.

Epitaxy

Epitaxy, is a transliteration of two Greek words *epi*, meaning "upon", and *taxis*, meaning "ordered". With respect to crystal growth it applies to the process of growing thin crystalline layers on a crystal substrate. In epitaxial growth, there is a precise crystal orientation of the film in relation to the substrate. The growth of epitaxial films can be done by a number of methods including molecular beam epitaxy, atomic layer epitaxy, and chemical vapour deposition, all of which will be described.

Epitaxy of the same material, such as a gallium arsenide film on a gallium arsenide substrate, is called homoepitaxy, while epitaxy where the film and substrate material are different is called heteroepitaxy.

Clearly, in homoepitaxy, the substrate and film will have the identical structure, however, in heteroepitaxy, it is important to employ where possible a substrate with the same structure and similar lattice parameters. For example, zinc selenide (zinc blende, a = 5.668 Å) is readily grown on gallium arsenide (zinc blende, a = 5.653 Å).

Alternatively, epitaxial crystal growth can occur where there exists a simple relationship between the structures of the substrate and crystal layer, such as is observed between Al_2O_3 (100) on Si (100). Whichever route is chosen a close match in the lattice parameters is required, otherwise, the strains induced by the lattice mismatch results in distortion of the film and formation of dislocations.

If the mismatch is significant epitaxial growth is not energetically Favourable, causing a textured film or polycrystalline untextured film to be grown. As a general rule of thumb, epitaxy can be achieved if the lattice parameters of the two materials are within about 5% of each other. For good quality epitaxy, this should be less than 1%. The larger the mismatch, the larger the strain in the film. As the film gets thicker and thicker, it will try to relieve the strain in the film, which could include the loss of epitaxy of the growth of dislocations.

It is important to note that the <100> directions of a film must be parallel to the <100> direction of the substrate. In some cases, such as Fe on MgO, the direction is parallel to the substrate [100]. The epitaxial relationship is specified by giving first the plane in the film that is parallel to the substrate.

ENERGY BANDS

Now that you have become reacquainted with matter and energy, we will continue our discussion with electron Behaviour. As stated earlier, orbiting electrons contain energy and are confined to definite energy levels. The various shells in an atom represent these levels.

Therefore, to move an electron from a lower shell to a higher shell a certain amount of energy is required. This energy can be in the form of electric fields, heat, light, and even bombardment by other particles. Failure to provide enough energy to the electron, even if the energy supplied is just short of the required amount, will cause it to remain at its present energy level.

Supplying more energy than is needed will only cause the electron to move to the next higher shell and the remaining energy will be wasted. In simple terms, energy is required in definite units to move electrons from one shell to the next higher shell.

These units are called Quanta (for example 1, 2, or 3 quanta). Electrons can also lose energy as well as receive it. When an electron loses energy, it moves to a lower shell. The lost energy, in some cases, appears as heat.

If a sufficient amount of energy is absorbed by an electron, it is possible for that electron to be completely removed from the influence of the atom. This is called Ionization. When an atom loses electrons or gains electrons in this process of electron exchange, it is said to be ionized.

For ionization to take place, there must be a transfer of energy that results in a change in the internal energy of the atom. An atom having more than its normal amount of electrons acquires a negative charge, and is called a Negative Ion. The atom that gives up some of its normal electrons is left with fewer negative charges than positive charges and is called a Positive Ion. Thus, we can define ionization as the process by which an atom loses or gains electrons.

Up to this point in our discussion, we have spoken only of isolated atoms. When atoms are spaced far enough apart, as in a gas, they have very little influence upon each other, and are very much like lone atoms.

But atoms within a solid have a marked effect upon each other. The forces that bind these atoms together greatly modify the Behaviour of the other electrons.

One consequence of this close proximity of atoms is to cause the individual energy levels of an atom to break up and form bands of energy.

Discrete (separate and complete) energy levels still exist within these energy bands, but there are many more energy levels than there were with the isolated atom. In some cases, energy levels will have disappeared.

Figure below shows the difference in the energy arrangement between an isolated atom and the atom in a solid. Notice that the isolated atom (such as in gas) has energy levels, whereas the atom in a solid has energy levels grouped into ENERGY BANDS.

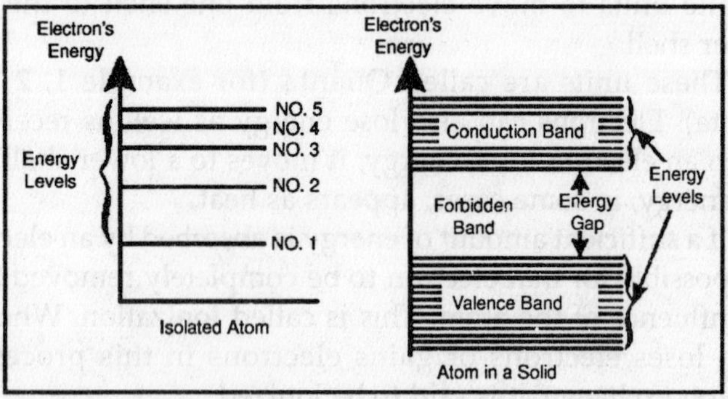

Fig. The Energy Arrangement in Atoms.

The upper band in the solid lines in figure above is called the CONDUCTION BAND because electrons in this band are easily removed by the application of external electric fields. Materials that have a large number of electrons in the conduction band act as good conductors of electricity.

Below the conduction band is the FORBIDDEN BAND or energy gap. Electrons are never found in this band, but may travel back and forth through it, provided they do not come to rest in the band.

The last band or VALENCE BAND is composed of a series of energy levels containing valence electrons. Electrons in this band are more tightly bound to the individual atom than the electrons in the conduction band.

However, the electrons in the valence band can still be moved to the conduction band with the application of energy, usually thermal energy.

There are more bands below the valence band, but they are not important to the understanding of semiconductor theory and will not be discussed.

The concept of energy bands is particularly important in classifying materials as conductors, semiconductors, and insulators. An electron can exist in either of two energy bands, the conduction band or the valence band.

All that is necessary to move an electron from the valence band to the conduction band so it can be used for electric current, is enough energy to carry the electron through the forbidden band.

The width of the forbidden band or the separation between the conduction and valence bands determines whether a substance is an insulator, semiconductor, or conductor. Figure uses energy level diagrams to show the difference between insulators, semiconductors, and conductors.

The energy diagram for the insulator shows the insulator with a very wide energy gap. The wider this gap, the greater the amount of energy required to move the electron from the valence band to the conduction band. Therefore, an insulator requires a large amount of energy to obtain a small amount of current. The insulator "insulates" because of the wide forbidden band or energy gap.

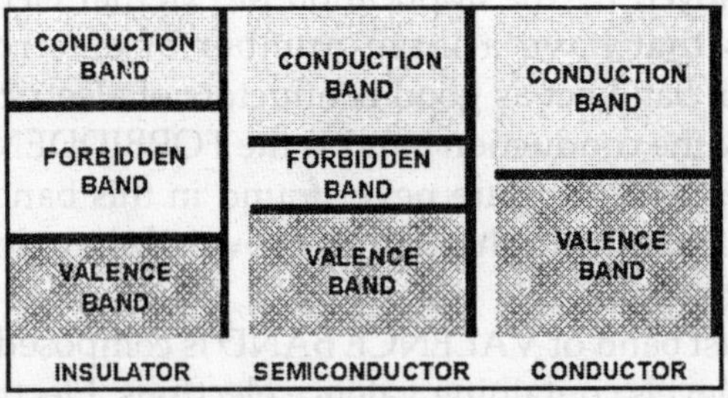

Fig. Energy Level Diagram.

The semiconductor, on the other hand, has a smaller forbidden band and requires less energy to move an electron from the valence band to the conduction band. Therefore, for a certain amount of applied voltage, more current will flow in the semiconductor than in the insulator.

The last energy level diagram in figure is that of a conductor. Notice, there is no forbidden band or energy gap and the valence and conduction bands overlap. With no energy gap, it takes a small amount of energy to move electrons into the conduction band; consequently, conductors pass electrons very easily.

Energy bands consisting of a large number of closely spaced energy levels exist in crystalline materials. The bands can be thought of as the collection of the individual energy levels of electrons surrounding each atom. The wavefunctions of the individual electrons, however, overlap with those of electrons confined to neighboring atoms. The Pauli exclusion principle does not allow the electron energy levels to be the same so that one obtains a set of closely spaced energy levels, forming an energy band. The energy band model is crucial to any detailed treatment of semiconductor devices. It provides the framework needed to understand the concept of an energy bandgap and that of conduction in an almost filled band as described by the empty states.

In this section, we present the free electron model and the Kronig-Penney model. Then we discuss the energy bands of semiconductors and present a simplified band diagram. We also introduce the concept of holes and the effective mass.

FREE ELECTRON MODEL

The free electron model of metals has been used to explain the photo-electric effect. This model assumes that electrons are free to move within the metal but are confined to the metal by potential barriers as illustrated by Figure below. The minimum energy needed to extract an electron from the metal equals q M, where M is the workfunction. This model is frequently used when analyzing metals. However, this model does not work well for semiconductors since the effect of the periodic potential due to the atoms in the crystal has been ignored.

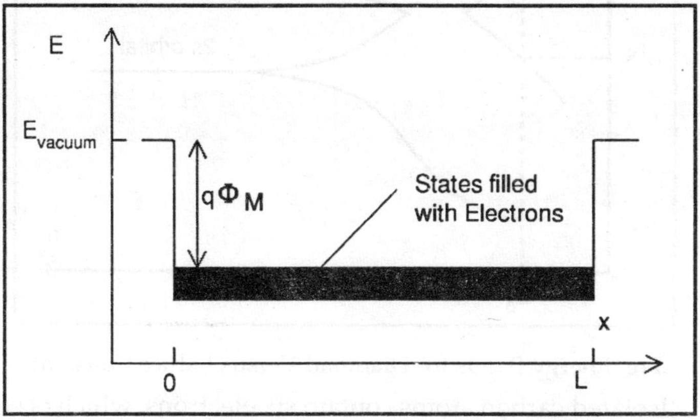

Fig. The Free Electron Model of a Metal.

ANALYSIS

The analysis of periodic potentials is required to find the energy levels in a semiconductor. This requires the use of periodic wave functions, called Bloch functions which are beyond the scope of this text. The result of this analysis is that the energy levels are grouped in bands, separated by energy band gaps. The Behaviour of electrons at the bottom of such a band is similar to that of a free electron. However, the electrons are affected by the presence of the periodic potential. The combined effect of the periodic potential is included by adjusting the value of the electron mass. This mass will be referred to as the effective mass.

The effect of a periodic arrangement on the electron

energy levels is illustrated by Figure. Shown are the energy levels of electrons in a carbon crystal with the atoms arranged in a diamond lattice. These energy levels are plotted as a function of the lattice constant, a.

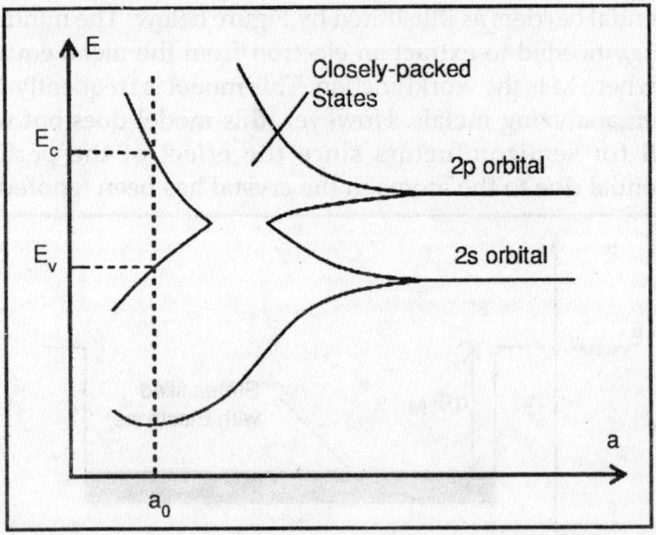

Fig. Energy Bands for Diamond Versus Lattice Constant.

Isolated carbon atoms contain six electrons, which occupy the 1s, 2s and 2p orbital in pairs. The energy of an electron occupying the 2s and 2p orbital is indicated on the figure. The energy of the 1s orbital is not shown. As the lattice constant is reduced, there is an overlap of the electron wavefunctions occupying adjacent atoms.

This leads to a splitting of the energy levels consistent with the Pauli exclusion principle. The splitting results in an energy band containing 2N states in the 2s band and 6N states in the 2p band, where N is the number of atoms in the crystal. A further reduction of the lattice constant causes the 2s and 2p energy bands to merge and split again into two bands containing 4N states each. At zero Kelvin, the lower band is completely filled with electrons and labeled as the valence band. The upper band is empty and labeled as the conduction band.

The Kronig-Penney model

The Kronig-Penney model demonstrates that a simple one-dimensional periodic potential yields energy bands as well as energy band gaps.

While it is an oversimplification of the three-dimensional potential and bandstructure in an actual semiconductor crystal, it is an instructive tool to demonstrate how the band structure can be calculated for a periodic potential, and how allowed and forbidden energies are obtained when solving the corresponding Schrödinger equation.

The potential assumed in the model is shown in the Figure below:

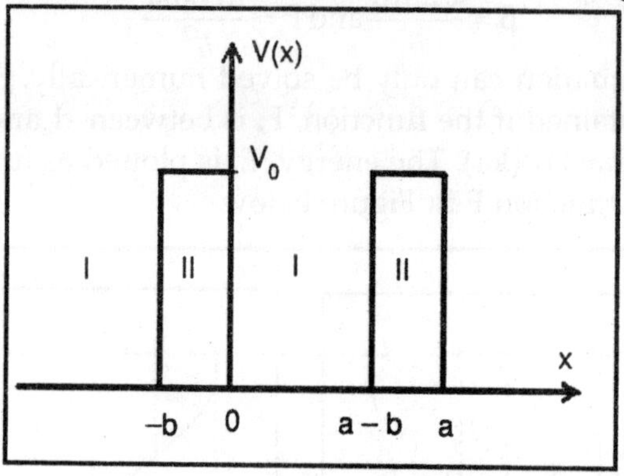

Fig. The Periodic Potential Assumed in the Kronig-Penney Model.

The potential barriers (region I) with width, b, are spaced by a distance (region II), a-b, and repeated with a period, a.

The Kronig-Penney consists of an infinite series of rectangular barriers with potential height, V0, and width, b, separated by a distance, a-b, resulting in a periodic potential with period, a.

The analysis requires the use of Bloch functions, traveling wave solutions multiplied with a periodic function, which has the same periodicity as the potential.

Solutions for k and E are obtained when the following equation is satisfied:

$$\text{Coska} = F = \frac{\alpha^2 - \beta^2}{2\alpha\beta} \sinh \beta(a - b) + \cosh \alpha b \cos \beta(a - b)$$

where,

$$\text{Coska} = F = P\frac{\sin\beta^2}{\beta a} + \cos\beta a$$

This transcendental equation can be further simplified for the case where the barrier is a delta function with area, V0b, for which it becomes:

$$\text{Coska} = F = P\frac{\sin\beta a}{\beta a} + \cos\beta a$$

with,

$$\beta = \frac{\sqrt{2mE}}{\hbar} \text{and}\, P = \frac{mV_0 ba}{\hbar^2}$$

This equation can only be solved numerically. Solutions are only obtained if the function, F, is between -1 and 1 since it has to equal cos(ka). The energy, E, is plotted as function of ka/and the function F in Figure below.

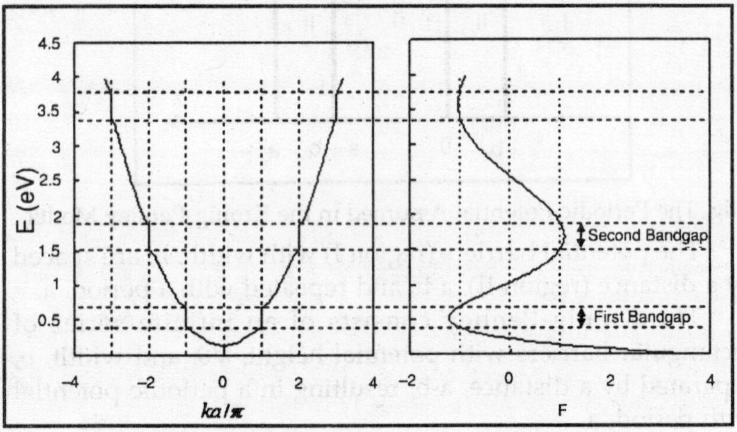

Fig. Graphical Solution to the Kronig-Penney model for a = 1nm and V0b = 0.2 nm-eV.

Shown is the energy, E, versus ka/and F, which has to equal cos(ka), from which one can identify the allowed energies.

The corresponding band structure is shown below (black curve) as well as the energy for a free electron (gray curve).

Three different forms are presented, namely the E(k) digram, the E(k diagram combined with the reduced-zone diagram as well as the reduced-zone diagram only.

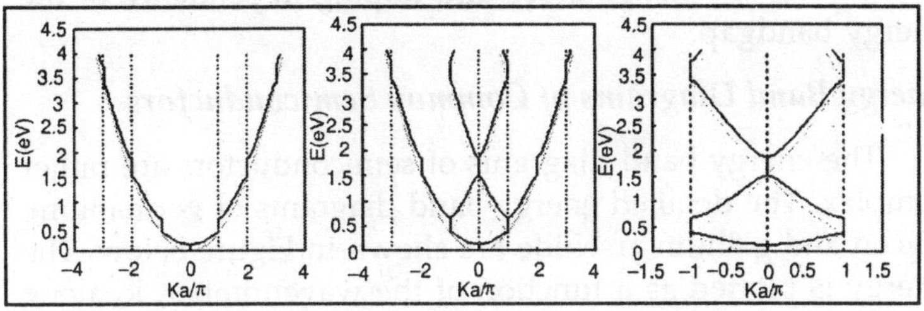

Fig. Energy Versus ka graph.

Shown are:

- The E(k) diagram,
- The E(k) diagram combined with the reduced-zone diagram
- The reduced-zone diagram only.

From Figure above we observe the following: The E(k) relation resembles a parabola except that only specific ranges of energies are valid solutions to Schrödinger's equation and therefore are allowed, while others are not. The range of energies for which there is no solution is referred to as an energy band gap.

The transitions between allowed and forbidden energies occur at non-zero integer multiples of ka/. These correspond to local minima and maxima of the E(k) relation. The reduced-zone diagram shown in Figure contains the first three bands and energy bandgaps. For instance the second energy bandgap occurs between 1.5 and 2 eV, between the band maximum of the second band and the band minimum of the third band.

- Energy bands of semiconductors
- Energy band diagrams of common semiconductors
- Simple energy band diagram of a semiconductor
- Temperature dependence of the energy bandgap
- Doping dependence of the energy bandgap

Complete energy band diagrams of semiconductors are very complex. However, most have features similar to that of the diamond crystal discussed.

In this section, we first take a closer look at the energy band diagrams of common semiconductors. We then present a simple diagram containing only the most important features and discuss the temperature and doping dependence of the energy bandgap.

Energy Band Diagrams of Common Semiconductors

The energy band diagrams of semiconductors are rather complex. The detailed energy band diagrams of germanium, silicon and gallium arsenide are shown in Figure below. The energy is plotted as a function of the wavenumber, k, along the main crystallographic directions in the crystal, since the band diagram depends on the direction in the crystal.

The energy band diagrams contain multiple completely-filled and completely-empty bands. In addition, there are multiple partially-filled band.

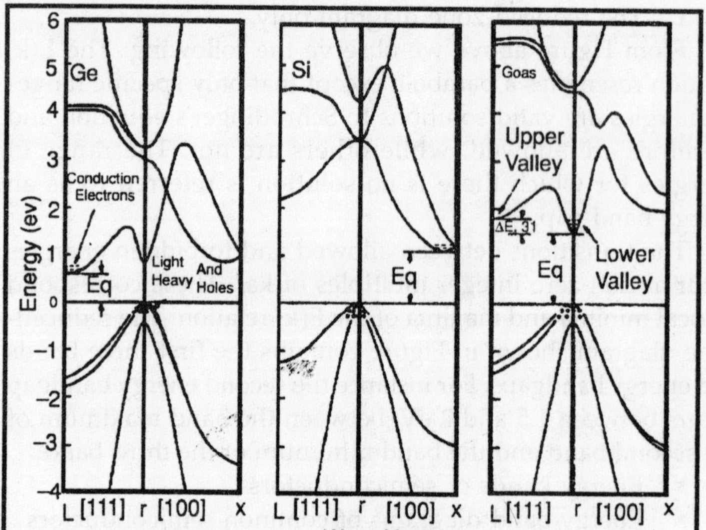

Fig. Energy Band Diagram of (a) Germanium, (b) Silicon and (c) Gallium Arsenide.

Fortunately, we can simplify the energy band diagram since only the electrons in the highest almost-filled band and the lowest almost-empty band dominate the Behaviour of the semiconductor. These bands are indicated on the figure by the + and - signs corresponding to the charge of the carriers in those bands.

Simple Energy Band Diagram of a Semiconductor

The energy band diagrams shown in the previous section are frequently simplified when analyzing semiconductor devices.

Since the electronic properties of a semiconductor are dominated by the highest partially empty band and the lowest partially filled band, it is often sufficient to only consider those bands. This leads to a simplified energy band diagram for semiconductors as shown in Figure below:

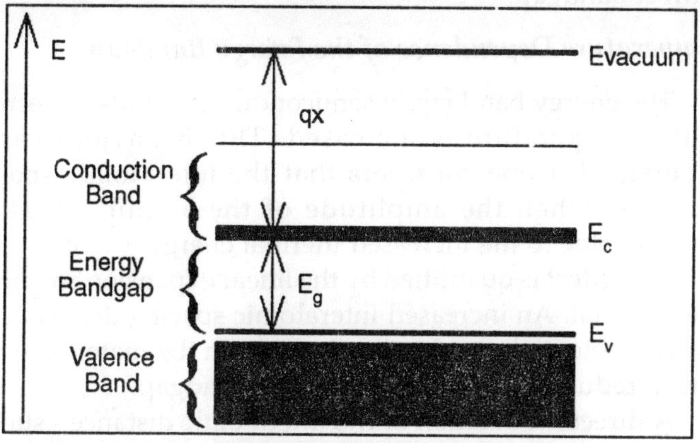

Fig. Energy Band Diagram to Describe Semiconductors.

Shown are the valence and conduction band as indicated by the valence band edge, Ev, and the conduction band edge, Ec. The vacuum level, Evacuum, and the electron affinity,, are also indicated on the figure.

The diagram identifies the almost-empty conduction band by a set of horizontal lines. The bottom line indicates the bottom edge of the conduction band and is labeled E_c. Similarly, the top of the valence band is indicated by a horizontal line labeled E_v.

The energy band gap, E_g, is located between the two

bands. The distance between the conduction band edge, Ec, and the energy of a free electron outside the crystal (called the vacuum level labeled Evacuum) is quantified by the electron affinity, multiplied with the electronic chargeq.

An important feature of an energy band diagram, which is not included on the simplified diagram, is whether the conduction band minimum and the valence band maximum occur at the same value for the wavenumber. If so, the energy bandgap is called direct.

If not, the energy bandgap is called indirect. This distinction is of interest for optoelectronic devices since direct bandgap materials provide more efficient absorption and emission of light. For instance, the smallest bandgap of germanium and silicon is indirect, while gallium arsenide has a direct bandgap.

Temperature Dependence of the Energy Bandgap

The energy bandgap of semiconductors tends to decrease as the temperature is increased. This Behaviour can be understood if one considers that the interatomic spacing increases when the amplitude of the atomic vibrations increases due to the increased thermal energy.

This effect is quantified by the linear expansion coefficient of a material. An increased interatomic spacing decreases the average potential seen by the electrons in the material, which in turn reduces the size of the energy bandgap.

A direct modulation of the interatomic distance - such as by applying compressive (tensile) stress - also causes an increase (decrease) of the bandgap. The temperature dependence of the energy bandgap, Eg, has been experimentally determined yielding the following expression for Eg as a function of the temperature, T:

$$E_g(T) = E_g(0) - \frac{\alpha T^2}{T + \beta}$$

Where Eg(0), and are the fitting parameters. These fitting parameters are listed for germanium, silicon and gallium arsenide in Table below:

Table. Parameters used to Calculate the Energy Bandgap of Germanium, Silicon and Gallium Arsenide (GaAs) as a Function of Temperature.

	Germanium	Silicon	GaAs
$E_g(0)$ (eV)	0.7437	1.166	1.519
α(meV/K)	0.473	0.473	0.541
β(K)	235	236	204

A plot of the resulting bandgap versus temperature is shown in Figure below for germanium, silicon and gallium arsenide.

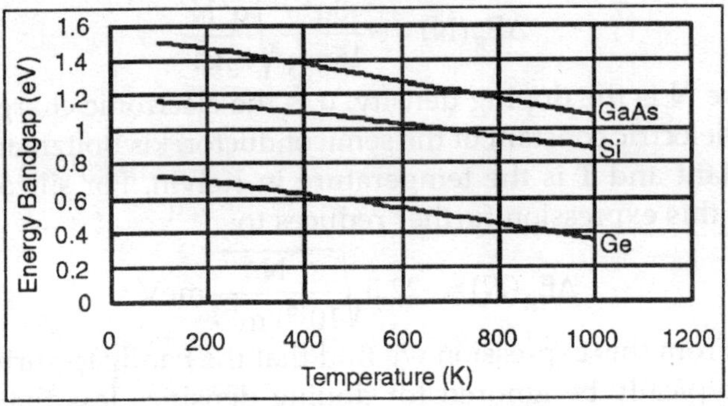

Fig. Temperature Dependence of the Energy Bandgap of Germanium (Ge), Silicon (Si) and Gallium Arsenide (GaAs).

Example Calculate the energy bandgap of germanium, silicon and gallium arsenide at 300, 400, 500 and 600 K.

Solution The bandgap of silicon at 300 K equals:

$$E_g(300K) = E_g(0K) - \frac{\alpha T^2}{T+\beta} = 1.166 - \frac{0.473 \times (300)^2}{300+636} = 1.12eV$$

Similarly one finds the energy bandgap for germanium and gallium arsenide, as well as at different temperatures, yielding:

	Germanium	Silicon	gallium Arsenide
T = 300K	0.66 eV	1.12 eV	1.42 eV
T = 400K	0.62 eV	1.09 eV	1.38 eV
T = 500K	0.58 eV	1.06 eV	1.33 eV
T = 600K	0.54 eV	1.03 eV	1.28 eV

Doping Dependence of the Energy Bandgap

High doping densities cause the bandgap to shrink. This effect is explained by the fact that the wavefunctions of the electrons bound to the impurity atoms start to overlap as the density of the impurities increase. For instance, at a doping density of 10^{18} cm^{-3}, the average distance between two impurities is only 10 nm. This overlap forces the energies to form an energy band rather than a discreet level. If the impurity level is shallow, this impurity band reduces the energy band of the host material by:

$$\Delta E_g(N) = -\frac{3q^2}{16\pi\varepsilon_3}\sqrt{\frac{q^2N}{e_3kT}}$$

where N is the doping density, q is the electronic charge, s is the dielectric constant of the semiconductor, kis Boltzmann'szz constant and T is the temperature in Kelvin. For silicon (r = 11.9) this expression further reduces to:

$$\Delta E_g(N) = -22.5\sqrt{\frac{N}{10^{18}\ cm^{-3}}}\text{meV}$$

From this expression we find that the bandgap shrinkage can typically be ignored for doping densities less than 10^{18} cm^{-3}. A plot of the change in bandgap energy with doping density is shown in Figure below:

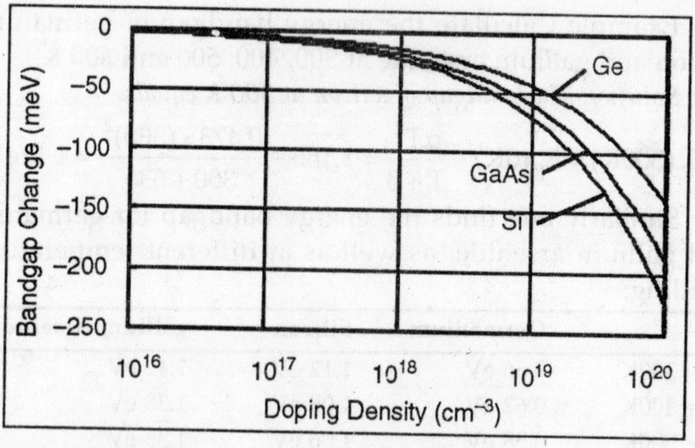

Fig. Doping Dependence of the Energy Bandgap of Germanium (Ge), Gallium Arsenide (GaAs), and Silicon (Si).

Metals, Insulators and Semiconductors

Once we know the bandstructure of a given material we still need to find out which energy levels are occupied and whether specific bands are empty, partially filled or completely filled.

Empty bands do not contain electrons. Therefore, they are not expected to contribute to the electrical conductivity of the material. Partially filled bands do contain electrons as well as available energy levels at slightly higher energies. These unoccupied energy levels enable carriers to gain energy when moving in an applied electric field. Electrons in a partially filled band therefore do contribute to the electrical conductivity of the material. Completely filled bands do contain plenty of electrons but do not contribute to the conductivity of the material. This is because the electrons cannot gain energy since all energy levels are already filled.

In order to find the filled and empty bands we must find out how many electrons can be placed in each band and how many electrons are available. Each band is formed due to the splitting of one or more atomic energy levels. Therefore, the minimum number of states in a band equals twice the number of atoms in the material. The reason for the factor of two is that every energy level can contain two electrons with opposite spin. To further simplify the analysis, we assume that only the valence electrons (the electrons in the outer shell) are of interest. The core electrons are tightly bound to the atom and are not allowed to freely move in the material.

Four different possible scenarios are shown in Figure below:

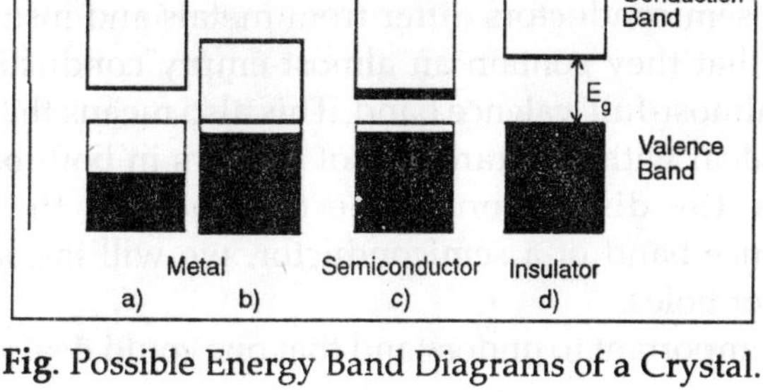

Fig. Possible Energy Band Diagrams of a Crystal.

Shown are:

a) A half filled band,
b) Two overlapping bands,
c) An almost full band separated by a small bandgap from an almost empty band
d) A full band and an empty band separated by a large bandgap.

A half-filled band is shown in above Figure. a). This situation occurs in materials consisting of atoms, which contain only one valence electron per atom. Most highly conducting metals including copper, gold and silver satisfy this condition.

Materials consisting of atoms that contain two valence electrons can still be highly conducting if the resulting filled band overlaps with an empty band.

This scenario is shown in b). No conduction is expected for scenario d) where a completely filled band is separated from the next higher empty band by a larger energy gap. Such materials behave as insulators.

Finally, scenario c) depicts the situation in a semiconductor. The completely filled band is now close enough to the next higher empty band that electrons can make it into the next higher band.

This yields an almost full band below an almost empty band. We will call the almost full band the valence band since it is occupied by valence electrons. The almost empty band will be called the conduction band, as electrons are free to move in this band and contribute to the conduction of the material.

Electrons and Holes in Semiconductors

The semiconductors differ from metals and insulators by the fact that they contain an"almost-empty"conduction band and an"almost-full"valence band. This also means that we will have to deal with the transport of carriers in both bands. To facilitate the discussion of the transport in the"almost-full"valence band of a semiconductor, we will introduce the concept of holes.

It is important to understand that one could deal with only

electrons if one is willing to keep track of all the electrons in the"almost-full"valence band. After all, electrons are the only real particles available in a semiconductor. The concepts of holes is introduced in semiconductors since it is easier to keep track of the missing electrons in an"almost-full"band, rather than keeping track of the actual electrons in that band. We will now first explain the concept of a hole and then point out how the hole concept simplifies the analysis.

Holes are missing electrons. They behave as particles with the same properties as the electrons would have when occupying the same states except that they carry a positive charge. This definition is illustrated further with Figure, which presents the energy band diagram in the presence of an electric field.

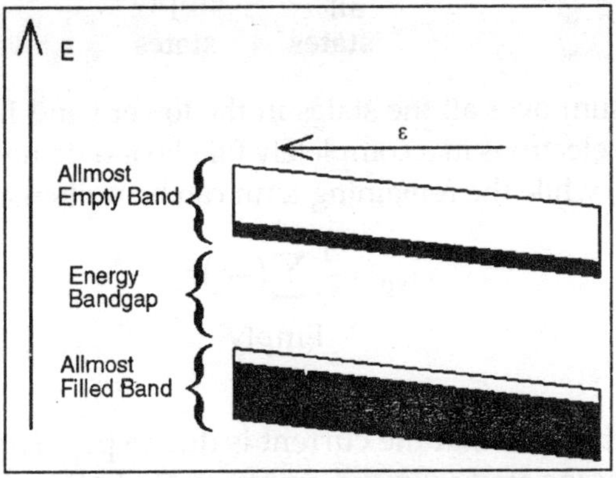

Fig. Energy Band Diagram in the Presence of a Uniform Electric Field.

Shown are the upper almost-empty band and the lower almost-filled band. The tilt of the bands is caused by an externally applied electric field. A uniform electric field is assumed which causes a constant gradient of the bands.

The electrons in the almost-empty band are negatively charged particles, which therefore move in a direction, which opposes the direction of the field. Electrons therefore move down hill in the upper band. Electrons in the lower band also move in the same direction. The total current density due to the electrons in the valence band can therefore be written as:

$$\text{Cos ka} = F = \frac{\alpha^2 - \beta^2}{2\alpha\beta} \sinh \alpha b \sin \beta(a-b) + \cosh \alpha b \cos \beta(a-b)$$

where V is the volume of the semiconductor, q is the electronic charge and v is the electron velocity. The sum is taken over all occupied or filled states in the lower band.

This equation can be reformulated by first taking the sum over all the states in the lower band and subtracting the current due to the electrons, which are missing in the almost-filled band. This last term therefore represents the sum taken over all the empty states in the lower band, or:

$$J_{vb} = \frac{1}{V}\left(\sum_{\substack{\text{all} \\ \text{states}}}(-q)v_i - \sum_{\substack{\text{empty} \\ \text{states}}}(-q)v_i\right)$$

The sum over all the states in the lower band has to equal zero since electrons in a completely filled band do not contribute to current, while the remaining term can be written as:

$$J_{vb} = \frac{1}{V}\sum_{\substack{\text{Empty} \\ \text{States}}}(+q)v_i$$

Which states that the current is due to positively charged particles associated with the empty states in the almost-filled band. We call these particles holes. Keep in mind that there is no real particle associated with a hole. Instead, the combined Behaviour of all the electrons, which occupy states in the almost-filled band, is the same as that of positively charge particles associated with the unoccupied states.

The reason the concept of holes simplifies the analysis is that the density of states function of a whole band can be rather complex. However, it can be dramatically simplified if only states close to the band edge need to be considered.

The effective mass concept

Electrons with an energy close to a band minimum behave

as free electrons, since the E-k relation can be approximated by a parabola. They accelerate in an applied electric field just like a free electron in vacuum. Their wavefunctions are periodic and extend over the size of the material.

The presence of the periodic potential, due to the atoms in the crystal without the valence electrons, changes the properties of the electrons. Therefore, the mass of the electron differs from the free electron mass, m_0.

Because of the anisotropy of the effective mass and the presence of multiple equivalent band minima, we define two types of effective mass:

- The effective mass for density of states calculations
- The effective mass for conductivity calculations.

The effective mass values for electrons and holes are listed together with the value of the smallest energy bandgap in Table below. Electrons in gallium arsenide have an isotropic effective mass so that the conductivity effective mass equals the density of states effective mass.

Table. Effective Mass of Carriers in Germanium, Silicon and Gallium Arsenide (GaAs)

		Germanium	Silicon	GaAs
Smallest energy Bandgap at 300K	E_g(eV)	0.66	1.12	1.424
Electron Effective Mass for Density of States Calculations	$\frac{m_e^* dos}{m_0}$	0.55	1.08	0.067
Hole Effective Mass for Density of States Calculations	$\frac{m_h^* dos}{m_0}$	0.37	0.811	0.45
Electron Effective Mass for Conductivity Calculations	$\frac{m_e^* cond}{m_0}$	0.12	0.26	0.067

Hole Effective Mass for Conductivity Calculations	$\frac{m_h^* cond}{m_0}$	0.21	0.386	0.34

The effective mass of a semiconductor is obtained by fitting the actual E-k diagram around the conduction band minimum or the valence band maximum by a paraboloid. While this concept is simple enough, the issue turns out to be substantially more complex due to the multitude and the occasional anisotropy of the minima and maxima. In this section we first describe the different relevant band minima and maxima, present the numeric values for germanium, silicon and gallium arsenide and introduce the effective mass for density of states calculations and the effective mass for conductivity calculations.

Most semiconductors can be described as having one band minimum at k = 0 as well as several equivalent anisotropic band minima at k 0. In addition there are three band maxima of interest close to the valence band edge.

Band Structure of Silicon

As an example we consider the band structure of silicon as shown in the figure below:

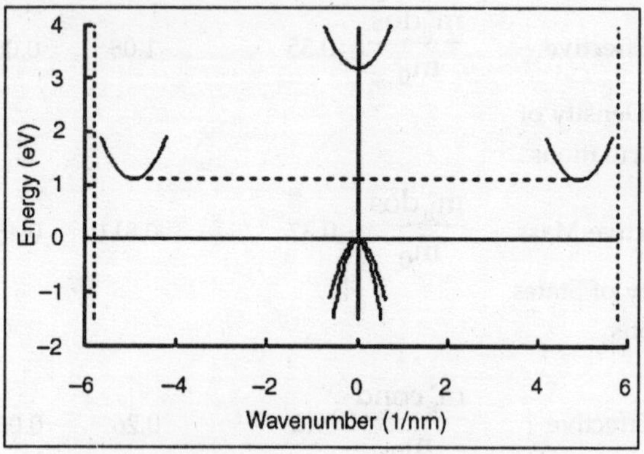

Fig. Simplified E-k Diagram of Silicon.

Shown is the E-k diagram within the first brillouin zone

and along the (100) direction. The energy is chosen to be to zero at the edge of the valence band. The lowest band minimum at k = 0 and still above the valence band edge occurs at Ec,direct = 3.2 eV.

This is not the lowest minimum above the valence band edge since there are also 6 equivalent minima at k = (x,0,0), (–x,0,0), (0,x,0), (0,–x,0), (0,0,x), and (0,0,–x) with x = 5 nm-1. The minimum energy of all these minima equals 1.12 eV = Ec,indirect.

The effective mass of these anisotropic minima is characterized by a longitudinal mass along the corresponding equivalent (100) direction and two transverse masses in the plane perpendicular to the longitudinal direction.

In silicon the longitudinal electron mass is me,l* = 0.98 m0 and the transverse electron masses are me,t* = 0.19 m0, where m0 = 9.11 × 10–31 kg is the free electron rest mass.

Two of the three band maxima occur at 0 eV. These bands are referred to as the light and heavy hole bands with a light hole mass of ml,h* = 0.16 m0 and a heavy hole mass of mh,h* = 0.46 m0.

In addition there is a split-off hole band with its maximum at Ev,so = –0.044 eV and a split-off hole mass of mh,so* = 0.29 m0. Effective mass and energy band minima and maxima of Ge, Si and GaAs

Effective Mass for Density of States Calculations

The effective mass for density of states calculations equals the mass which provides the density of states using the expression for one isotropic maximum or minimum or:

$$g_c(E) = \frac{8\pi\sqrt{2}}{h^3} m_e^{3/2} \sqrt{E - E_c} \text{ for } E \geq E_c$$

for the density of states in the conduction band and:

$$g_v(E) = \frac{8\pi\sqrt{2}}{h^3} m_e^{3/2} \sqrt{E_v - E} \text{ for } E \geq E_v$$

for the density of states in the valence band. For instance, for a single band minimum described by a longitudinal mass and two transverse masses the effective mass for density of states calculations is the geometric mean of the three masses. Including the fact that there are several equivalent minima at the same energy one obtains the effective mass for density of states calculations from:

$$m^*_{e,dos} M_c^{2/3} \sqrt[3]{m_l m_t m_t}$$

Where Mc is the number of equivalent band minima. For silicon one obtains:

$$m^*_{e,dos} M_c^{2/3} \sqrt[3]{m_l m_t m_t} = 6^{2/3} \sqrt[3]{0.98 \times 0.19 \times 019 m_0} = 1.0 m_0$$

Effective Mass for Conductivity Calculations

The effective mass for conductivity calculation is the mass, which is used in conduction related problems accounting for the detailed structure of the semiconductor. These calculations include mobility and diffusion constants calculations. Another example is the calculation of the shallow impurity levels using a hydrogen-like model.

As the conductivity of a material is inversionally proportional to the effective masses, one finds that the conductivity due to multiple band maxima or minima is proportional to the sum of the inverse of the individual masses, multiplied by the density of carriers in each band, as each maximum or minimum adds to the overall conductivity.

For anisotropic minima containing one longitudinal and two transverse effective masses one has to sum over the effective masses in the different minima along the equivalent directions.

The resulting effective mass for bands, which have ellipsoidal constant energy surfaces, is given by:

$$m^*_{e,cond} = \frac{3}{\frac{1}{m_l} + \frac{1}{m_t} + \frac{1}{m_t}}$$

Provided the material has an isotropic conductivity as is the case for cubic materials. For instance electrons in the X minima of silicon have an effective conductivity mass given by:

$$m^*_{e,cond} = \frac{3}{\frac{1}{m_l}+\frac{1}{m_t}+\frac{1}{m_t}} = \frac{3}{\frac{1}{0.89}+\frac{1}{0.19}+\frac{1}{0.19}} = 0.26m_0$$

Derivation of the Kronig-Penney equation

The solution to Schrödinger's equation for the Kronig-Penney potential previously shown in Figure and discussed is obtained by assuming that the solution is a Bloch function, namely a traveling wave solution of the form, eikx, multiplied with a periodic solution, u(x), which has the same periodicity as the periodic potential.

The total wavefunction is therefore of the form:

$$\psi(x) = u(x)e^{ikx}$$

where u(x) is the periodic function as defined by u(x) = u(x + a), and k(x) is the wave number. We now rewrite Schrödinger equation using this wave function considering first region I, between the barriers where V(x) = 0 and then region II, the barrier region where V(x) = V0:

In region I, Schrödinger's equation becomes:

$$\frac{d^2u_1(x)}{dx^2} + 2ik\frac{du_1(x)}{dx} + (\beta^2 - k^2)u_1(x) = 0 \text{for} 0 < x < a - b$$

With,

$$\beta = \frac{\sqrt{2mE}}{\hbar}$$

While in region II, it becomes:

$$\frac{d^2u_{II}(x)}{dx^2} + 2ik\frac{du_{II}(x)}{dx} - (k^2 + \alpha^2)u_{II}(x) = 0 \text{for} a - b < x - a$$

With,

$$\alpha = \frac{\sqrt{2m(V_0 - E)}}{\hbar}$$

The solution to equations are of the form:

$$u_1(x) = (A\cos\beta x + B\sin\beta x)e^{-ikx} \text{for} 0 < x < a - b$$

$$u_{II}(x) = (C\cosh\alpha x + D\sin\alpha x)e^{-ikx} \text{for } a-b<x<a$$

Since the potential, V(x), is finite everywhere, the solutions for uI(x) and uII(x) must be continuous as well as their first derivatives. Continuity at x = 0 results in:

$$u_I(0) = u_{II}(0) \text{ so that } A = C$$

and continuity at x = a-b combined with the requirement that u(x) be periodic:

$$u_I(a-b) = u_{II}(-b)$$

results in:

$$A\cos\beta(a-b) + B\sin\beta(a-b))e^{-ik(a-b)} =$$

$$(C\cos\alpha b - D\sinh\alpha b)e^{ikb}$$

Continuity of the first derivative at x = 0 requires that:

$$\left.\frac{du_1(x)}{dx}\right|_{x=0} = \left.\frac{du_{II}(x)}{dx}\right|_{x=0}$$

The first derivatives of uI(x) and uII(x) are:

$$\frac{du_1(x)}{dx} = (-A\beta\sin\beta x + B\beta\cos\beta x)e^{-ikx}$$

$$ik(A\cos\beta x + B\sin\beta x)e^{-ikx}$$

$$\frac{du_{II}(x)}{dx} = (-C\alpha\sinh\alpha x + B\beta\cos\beta x)e^{-ikx}$$

$$-ik(A\cos\beta x + B\sin\beta x)e^{-ikx}$$

So that becomes:

$$B\beta - ikA = D\alpha - ikC$$

Finally, continuity of the first derivative at x = a–b, again combined with the requirement that u(x) is periodic, results in:

$$\left.\frac{du_I(x)}{dx}\right|_{x=a-b} = \left.\frac{du_{II}(x)}{dx}\right|_{x=-b}$$

so that,

$$(-A\beta\sin\beta(a-b) + B\beta\cos\beta(a-b))e^{-ik(a-b)}$$

$$-ik(A\cos\beta(a-b) + B\sin\beta(a-b))e^{-ik(a-b}$$

$$= (-C\alpha\sinh\alpha b + D\alpha\cosh\alpha b)e^{ikb} - ik(C\cosh\alpha b - D\sinh\alpha b)e^{ikb}$$

This equation can be simplified using equation shown to:

$$(-A\beta\sin\beta(a-b)+B\beta\cos\beta(a-b))=$$

$$(-C\alpha\sinh\alpha b+D\alpha\cosh\alpha b)e^{ika}$$

As a result we have four homogenous equations shown above with four unknowns, A, B, C, and D, for which there will be a solution if the determinant of this set of equations is zero, or:

$$\begin{vmatrix} 1 & 0 & -1 & 0 \\ 0 & \beta & 0 & -\alpha \\ \cos\beta(a-b) & \sin\beta(a-b) & -\cosh\alpha b\exp ika & \sinh\alpha b\exp ika \\ -\beta\sin\beta(a-b) & \beta\cos\beta(a-b) & \alpha\sinh\alpha b\exp ika & -\alpha\cosh\alpha b\exp ika \end{vmatrix}=0$$

The first row of the determinant represents equation, the second row is obtained by combining above equations the third row represents equation and the fourth row represents equation as shown above.

This determinant can be rewritten as two determinants, each with three rows and columns, while replacing cos β(a–b) by β_c, sin β(a–b) by β_s, cosh αb eika by α_c and sinh αb e^{ika} by α_s, which results in:

$$\begin{vmatrix} \beta & 0 & -\alpha \\ \beta_s & \alpha_c & \alpha_s \\ \beta\beta_c & \alpha\alpha_s & -\alpha\alpha_c \end{vmatrix}=\begin{vmatrix} 0 & \beta & -\alpha \\ \beta_c & \beta_s & \alpha_s \\ -\beta\beta_s & \beta\beta_c & -\alpha\alpha_c \end{vmatrix}$$

Working out the determinants and using $\beta_c^2+\beta_s^2=1$, and,

$$\alpha_s^2+\alpha_c^2=e^{2ika},$$

one finds:

$$\left(\alpha^2-\beta^2\right)\alpha_s\beta_s+2\alpha\beta\alpha_c\beta=\alpha\beta\left(1+e^{2ika}\right)$$

And finally, substituting c, s, c and s:

$$\cos ka=F=\frac{\alpha^2\beta^2}{2a\beta}\sinh\alpha b\sin\beta(a-b)+\cosh\alpha b\cos\beta(a-b)$$

where eika + e-ika was replaced by 2coska. Since the result is independent of the sign of k, this equation also covers all solutions obtained when replacing e-ik by eik in equations. A further simplification is obtained as the barrier width, b, is reduced to zero while the barrier height, V_0, is increased to infinity in such manner that the product, bV_0, remains constant and the potential becomes a delta function train at x = a and repeated with a period of a, namely bV_0 (x - b - na) where n is an integer. As b approaches zero, sinh b approaches b.

Equation as shown below reduces to:

$$\cos ka = F = P\frac{\sin \beta a}{\beta a} + \cos \beta a$$

With,

$$\beta = \frac{\sqrt{2mE}}{\hbar} \text{ and } P = \frac{mV_0 ba}{\hbar^2}$$

Alternate Derivation of the Kronig-Penney Equation

First we recognize that the Schrödinger equation is of the form:

$$\frac{d^2\psi}{dx^2} + U(x)\psi = 0, \text{ where } U(x) = \frac{2m(V(x) - E)}{\hbar^2}$$

All solutions to this equation can be written as a linear combination of two arbitrary, but linearly independent, particular solutions. These can be chosen to be real since U(x) is a real function. Let the two particular solutions be $\Phi_1(x)$ and $\Phi_2(x)$.

From the periodicity of U(x):

$$U(x + a) = U(x)$$

it follows that 1(x+a) and 2(x+a) are also solutions, so that there exist real coefficients A_{11}, A_{12}, A_{21} and A_{22} for which:

$$\begin{bmatrix} \psi_1(x+a) \\ \psi_2(x+a) \end{bmatrix} = \begin{bmatrix} A_{11} & A_{12} \\ A_{21} & A_{22} \end{bmatrix} \times \begin{bmatrix} \psi_1(x) \\ \psi_2(x) \end{bmatrix}$$

The solution for the actual wavefunction $\Phi(x)$ is a linear combination of the real solutions, $\Phi_1(x)$ and $\Phi_2(x)$, with coefficients p and q:

$$\psi(x) = p\psi_1(x) + q\psi_2(x)$$

such that,

$$\psi(x + a) = \lambda\psi(x)$$

This is an eigenvalue problem and a solution exists for,:

$$\begin{bmatrix} A_{11} - \lambda & A_{12} \\ A_{21} & A_{22} - \lambda \end{bmatrix} = 0$$

The corresponding eigenvalues are:

$$\lambda_{1,2} = \frac{A_{11} + A_{22}}{2} \pm i\sqrt{(A_{11}A_{22} - A_{12}A_{21}) - \left(\frac{A_{11} + A_{22}}{2}\right)^2}$$

Since the wavefunction Φ (x) can also be expressed using Bloch functions:

$$\psi(x) = u(x)\exp(\pm ikx) \text{ with } u(x + a) = u(x)$$

The eigenvalue equation can be reformulated as:

$$\psi(x + a) = u(x)\exp(\pm ika)\exp(\pm ikx) = \psi(x)\exp(\pm ika)$$

so that,

$$\lambda_{1,2} = \exp(\pm ika)$$

Combining above equations results in:

$$A_{11} + A_{22} = \lambda_1 + \lambda_2 = 2\cos ka$$

We now construct the real functions $\Phi_1(X)$ and $\Phi_2(X)$ corresponding to the Kronig-Penney potential, while requiring the functions and their derivatives to be continuous at x = 0:

$$\psi_1(x) = \exp ax, \text{for} - b < x < 0$$

$$\psi_1(x) = \cos\beta x + \frac{\alpha}{\beta}\sin + \beta x, \text{for} 0 < x < a - b$$

and,

$$\psi_2(x) = \exp(-ax), \text{for} - b < x < 0$$

$$\psi_1(x) = \cos\beta x + \frac{\alpha}{\beta}\sin + \beta x, \text{for} 0 < x < a - b$$

Where,

$$\alpha = \frac{\sqrt{2m(V_0 - E)}}{\hbar} \text{and} \beta = \frac{\sqrt{2mE}}{\hbar}$$

The wavefunctions in the section between x = a-b and x = a, are now obtained.

$$\psi_1(x) = A_{11}\exp\alpha x + A_{12}\exp(-\alpha x), \text{for} a - b < x < a$$

$$\psi_2(x) A_{21}\exp\alpha x + A_{22}\exp(-\alpha x), \text{for} a - b < x < a$$

Both functions and their derivatives also need to be continuous at x = a–b, leading to:

$$A_{11} = \frac{\alpha^2 - \beta^2}{2\alpha\beta} \exp \alpha b \sin \beta(a-b) + \exp \alpha b \cos \beta(a-b)$$

$$A_{22} = \frac{\alpha^2 - \beta^2}{2\alpha\beta} \exp(-\alpha b) \sin \beta(a-b) + \exp(-\alpha b) \cos \beta(a-b)$$

ENERGY GAP

Valence band and the conduction band lies in the energy gap. According to the Bohr's theory, free electrons in an isolated atom have certain definite discrete amount of energy. If large number of atoms are brought close to one another to form a crystal, they begin to influence each other.

The valence electrons are attracted by the nucleus of the other atoms. This brings about a considerable modification in the case of energy levels of the electrons in the outer shells.

The process of splitting of energy levels can be understood as follows:

- If interatomic spacing of atoms is very large i.e., r = d>>a, there is no interatomic separation. Each atom in the crystal behaves as free atom. Take for example silicon whose electronic configuration is $1s^2\ 2s^2\ 2p^6\ 3s^2\ 3p^2$. If N atoms were to be considered in silicon crystal, then there will be 2N electrons filling 2N possible energy levels in 3s, 6N possible levels in 3p of which only 2N is completely filled.
- When the spacing is progressively decreased i.e., c<r<d, there is no visible splitting of energy levels.
- When r = c, the 3s and 3p electrons of neighbouring silicon atoms becomes appreciable. The energy of electrons of each atom starts changing, whereas the energies of electrons in the inner shell do not change.
- When r lies between b and c, the energy levels get slightly changed and instead of a single 3s or 3p levels, we get a large number of closely packed levels. This collection of closely spaced energy levels is called an energy band.

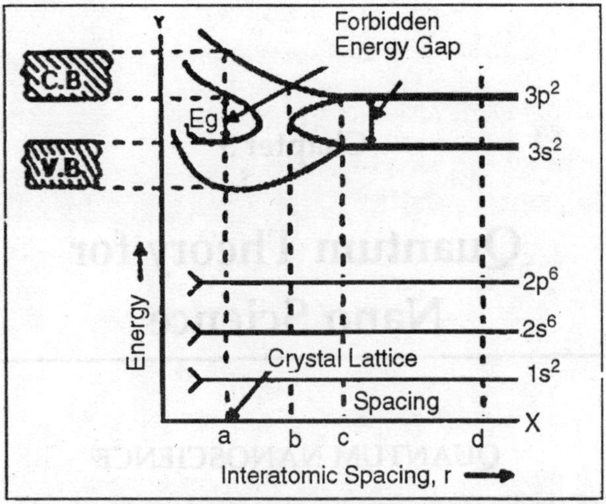

- When r = b>a, the gap between 3s and 3p completely disappear and the 8N energy levels are (2N of 3s and 6N of 3p sub shells) continuously distributed. In this stage 4N levels are filled and 4N levels are empty.
- When r = a i.e., actual spacing in the crystal the 4N filled energy levels are separated from 4N unfilled energy levels. This gap or separation is called the forbidden gap. E.g., the lower completely filled band is called valence band and upper ur.filled band is called conduction band.

Chapter 3

Quantum Theory for Nano Science

QUANTUM NANOSCIENCE

Nanotechnology can be classified under the headings: materials and devices. Nanomaterials are a suit of technologies, techniques and processes that seek to create new materials by manipulating the molecular and macro molecular constituents of matter. Nano devices require more complex processing of constituents at the nanoscale and typically use diverse materials, nanofabrication and multiple layers and even moving parts. An example of materials nanotechnology is the chemical synthesis of semiconductor nanocrystal quantum dots. These materials have optical properties governed by the quantisation of the energies of electrons in the crystal. They are being applied to new kinds of optical filters and display technologies.

An example of devices nanotechnology is seen in the global race to build a quantum computer. New and complex nanofabrication methods are being developed to build key functional elements of a solid state quantum computer using atomically precise positioning of phosphorous in silicon with nanometere scale electrical addressing. Both materials and devices can exhibit new features and functions determined by the rules of the quantum world, rules that govern at the nanoscale.

Quantum nanoscience is the application of quantum theory to the design of new nanoscale materials and devices.

Quantum Nanoscience explains functionality and structure in natural or engineered nanoscale systems through quantum mechanisms such as discretisation, superposition and entanglement.

In the 19th century, decades of practice with heat engines led to the new science of thermodynamics. The understanding of the world captured by thermodynamics is now part of the fabric of engineering and effective design across a vast range of different technologies. Like thermodynamics, quantum nanoscience is an enabling science for engineering and design of new nanotechnologies.

The Oliphant Conference on Quantum Nanoscience is one of the first international conferences bringing together physicists, chemists and engineers to develop common principles as a foundation for a future quantum nanotechnology.

The resulting understanding and systematisation will engender new materials and devices that cannot be reached by incremental progress. The purpose of quantum nanoscience is to enable generation-after-next nanotechno-logies, new industries and ultimately new economies.

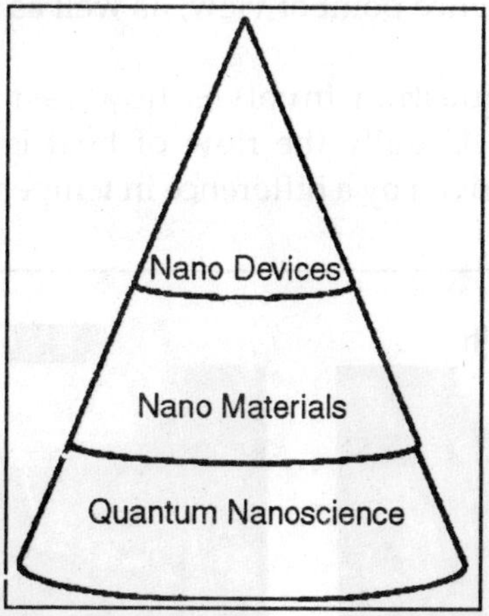

NANO MATERIALS

Margaret Murnane and Henry Kapteyn are developing

compact coherent (i.e., laserlike) X-ray sources that have pulse durations of femtoseconds to attoseconds (10^{-15} to 10^{-18} seconds). The short wavelength of X-rays combined with the presence of characteristic X-ray absorption edges in molecules and materials makes them an ideal probe of the nanoworld. There they have the ability to probe ultrafast element- and chemical-specific dynamics in real time and in thick samples.

Applications include monitoring switching times in molecular electronics, understanding the fastest dynamics in magnetism when light interacts coherently with spins, developing super-resolution microscopy, monitoring the first steps in catalysis, and understanding how heat flows in nanostructures.

The new probes, which complement existing nanoprobes, such as atomic-force microscopes, could become indispensable in the quest to develop practical nanoscale"machines."They are also expected to aid our understanding of nanoscience and nanotechnology at the level of individual electrons, atoms, and molecules.

Recently, the Kapteyn/Murnane group has explored several topics in nanoscience that are of interest from a fundamental science point of view, as well as being important technologically.

The first question involves how fast heat moves in nanostructures. Usually the flow of heat is determined by diffusion, and driven by a difference in temperature or thermal gradient.

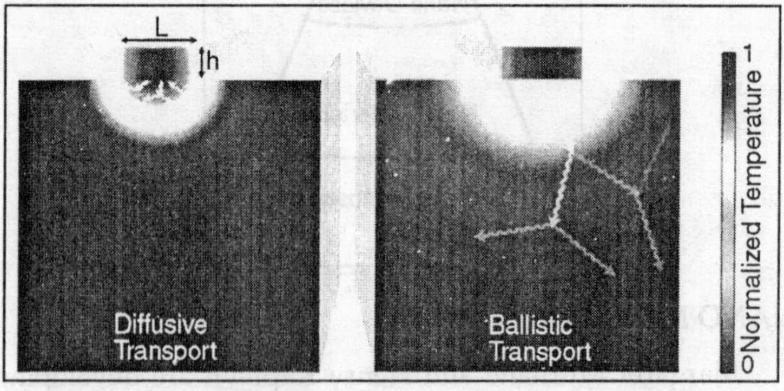

Illustration of the difference between diffusive and ballistic thermal transport across an interface. The Fourier diffusive heat equation describes diffusive heat transfer (left). In quasi-ballistic transport(right), the Fourier diffusive heat equation breaks down.

Credit: Mark Siemens

In this regime, it is not even possible to define a temperature. By heating an array of nanowires with a femtosecond laser and measuring how fast heat dissipates into two different materials, the researchers were able to observe and accurately measure ballistic thermal transport from a tiny nanoscale hotspot for the first time.

Ballistic heat transport is driven by point-to-point transport of energy quanta. These measurements could have a significant impact on heat management and reliability in future nanoscale devices.

In other exciting work, the Kapteyn/Murnane group is exploring"lensless"diffractive imaging, which enables scientists to use coherent (laserlike and directed) X-ray beams to take pictures of an object. The researchers shine a beam on the object and collect the diffracted photons. Sophisticated computational algorithms then process the diffraction pattern to reveal the image. The most rapid natural events occur at the attosecond (10^{-18} s) or femtosecond (10^{-15} s) time scales; therefore they must occur in the nanoworld, because light travels a distance of only 300 nm in 1 fs. Because the spatial resolution of any camera or microscope is limited by the wavelength of the illuminating light, short bursts of light at very small wavelengths are needed to visualize ultrafast motion directly at the nanoscale.

Pulses of X-rays are ideal for directly visualizing how the electrons that bind molecules together adjust as a molecule's structure changes during chemical reactions or conformational changes; how the motion of a biomolecule relates to its function; how fast magnetic materials can switch state in data storage devices; or how light can be used to coherently interact with, and control, the electronic properties of materials.

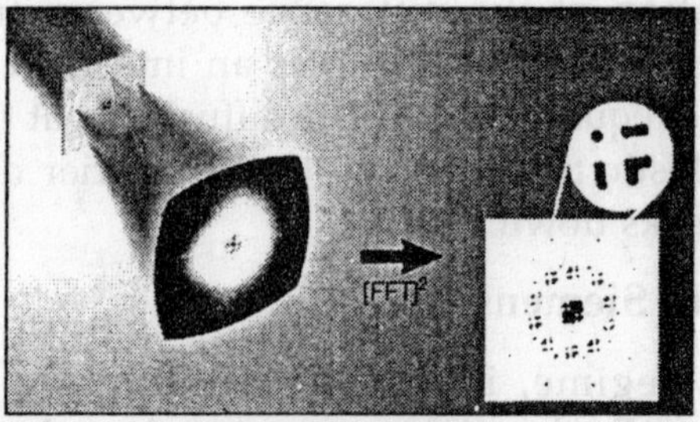

Fig. Soft X-ray Diffraction Patterns and Reconstructed Images Made with Lenless Coherent Imaging.

Finally, the Kapteyn/Murnane group is investigating some of the fastest processes in our natural world - the dizzying dance of electrons in materials and on surfaces. Using a laser-assisted version of the original photoelectric effect (Lape) discovered by Einstein, the scientists have been able to follow how fast an electron can move from one state to another at a surface. Interesting questions to be answered include how to make catalysts and photovoltaics more efficient.

The concept of a Universal Computer, a device able to compute anything computable, dates back to Babbage in the early part of the 19th century, and to Turing, Church, and von Neumann in the 20th century.

The concept of a "Universal Constructor" is perhaps more recent. The concept was well understood by von Neumann in the 1940's, who defined a "Universal Constructor" in a two-dimensional cellular automata world.

Such a model is a mathematical abstraction something like an infinite checkerboard, with several different types of "checkers" that might be on each square. The different pieces spontanesouly change and move about depending on what pieces occupy neighboring squares, in accordance with a pre-defined set of rules.

Von Neumann used the concept of a Universal Constructor in conjunction with a Universal Computer as the core components in a self-replicating system. The possibility of fabricating structures by putting "...the atoms down where the chemist says..." was recognized by Feynman in 1959.

Drexler recognized the value of the "assembler" in 1977. The assembler is analogous to von Neumann's Universal Constructor, but operates in the normal three dimensional world and can build a large, atomically precise structures by manipulating atoms and small clusters of atoms. He published the concept in 1981 and in subsequent work has proposed increasingly detailed designs for such devices.

The basic design of Drexler's assembler consists of:

- A molecular computer,
- One or more molecular positioning devices (which might resemble very small robotic arms),
- A well defined set of chemical reactions (perhaps one or two dozen) that take place at the tip of the arm and are able to fabricate a wide range of structures using site-specific chemical reactions.

It is now common for forecasts of future technical abilities to include the ability to fabricate molecular devices and molecular machines with atomic precision. While there continues to be debate about the exact time frame, it is becoming increasingly accepted that we will, eventually, develop the ability to economically fabricate a truly wide range of structures with atomic precision.

This will be of major economic value. Most obviously a molecular manufacturing capability will be a prerequisite to the construction of molecular logic devices. The continuation of present trends in computer hardware depends on the ability to fabricate ever smaller and ever more precise logic devices at ever decreasing costs.

The limit of this trend is the ability to fabricate molecular logic devices and to connect them in complex patterns at the molecular level. The manufacturing technology needed will, almost of necessity, be able to economically manufacture large structures (computers) with atomic precision (molecular logic elements). This capability will also permit the economical manufacture of materials with properties that border on the limits imposed by natural law. The strength of materials, in particular, will approach or even exceed that of diamond. Given the broad range of manufactured products that devote substantial mass to load-bearing members, such a development by itself will have a significant impact.

A broad range of other manufactured products will also benefit from a manufacturing process that offers atomic precision at low cost. Given the promise of such remarkably high payoffs it is natural to ask exactly what such systems will look like, exactly how they will work, and exactly how we will go about building them.

One might also enquire as to the reasons for confidence that such an enterprise is feasible, and why one should further expect that our current understanding of chemistry and physics (embodied in a number of computational chemistry packages) should be sufficient to explain the operating principles of such systems.

It is here that the value of computational nanotechnology can be most clearly seen. Molecular machine proposals, provided that they are specified in atomic detail (and are of a size that can be dealt with by current software and hardware), can be modeled using the tools of computational chemistry.

There are two modeling techniques of particular utility. The first is molecular mechanics, which utilizes empirical force fields to model the forces acting between nuclei. The second is higher order ab initio calculations, which will be discussed in a few paragraphs.

A few links to internet computational chemistry resources are provided here to provide those new to the area with some feel for what's available.

MOLECULAR MECHANICS

Molecular mechanics allows computational modeling of the positions and trajectories of the nuclei of individual atoms without an undue computational load. Current packages available on personal computers can readily do energy minimizations on systems with thousands of atoms, while supercomputers can handle systems with hundreds of thousands of atoms or more.

More complex analyses, particularly analyses that involve searching through large configuration spaces, can limit the size of system that can be effectively handled. As will be discussed later the need to search through large configuration spaces (as when determining the native folded structure of an arbitrary protein) can be avoided by the use of relatively rigid structures (which differ from relatively floppy proteins and have few possible configurations).

The modeling of machine components in vacuum reduces the need to model solvation effects, which can also involve significant computational effort. In molecular mechanics the individual nuclei are usually treated as point masses. While quantum mechanics dictates that there must be a certain degree of positional uncertainty associated with each nucleus, this positional uncertainty is normally significantly smaller than the typical internuclear distance. Bowen and Allinger[16] provide a good recent overview of the subject.

While the nuclei can reasonably be approximated as point masses, the electron cloud must be dealt with in quantum mechanical terms. However, if we are content to know only the positions of the nuclei and are willing to forego a detailed understanding of the electronic structure, then we can effectively eliminate the quantum mechanics.

For example, the H_2 molecule involves two nuclei. While it would be possible to solve Schrodinger's equation to determine the wave function for the electrons, if we are content simply to know the potential energy contributed by the electrons (and do not enquire about the electron distribution) then we need only know the electronic energy as a function of the distance between the nuclei.

That is, in many systems the only significant impact that the electrons have on nuclear position is to make a contribution to the potential energy E of the system. In the case of H_2, E is a simple function of the internuclear distance r. The function E(r) summarizes and replaces the more complex and more difficult to determine wave function for the electrons, as well as taking into account the inter-nuclear repulsion and the interactions between the electrons and the nuclei.

The two hydrogen nuclei will adopt a position that minimizes E(r). As r becomes larger, the potential energy of the system increases and the nuclei experience a restoring force that returns them to their original distance. Similarly, as r becomes smaller and the two nuclei are pushed closer together, we also find that a restoring force pushes them farther apart, again restoring them to an equilibrium distance.

More generally, if we know the positions r_1, r_2.... r_N of N nuclei, then $E(r_1, r_2 .... r_N)$ gives the potential energy of the system. Knowing the potential energy as a function of the nuclear positions, we can readily determine the forces acting on the individual nuclei and therefore can compute the evolution of their position over time.

The function E is a newtonian potential energy function (not quantum mechanical), despite the fact that the particular value of E at a particular point could be computed from Schrodinger's equation.

That is, the potential energy E is a newtonian concept, but the particular values of E at particular points are determined by Schrodinger's equation. Often called the Born-Oppenheimer approximation this approach allows a great conceptual and practical simplification in the modeling of molecular systems. While it would in principle be possible to determine E by solving Schrodinger's equation, in practice it is usual to use available experimental data and to infer the nature and shape of E by interpolation.

This approach, in which empirically derived potential energy functions are created by interpolation from experimental data, has spawned a wide range of potential energy functions, many of which are sold commercially. Because the gradient of the potential energy function E defines a conservative force field F, molecular mechanics methods are also called "force field" methods.

While it is common to refer to "empirical force field" methods, the more recent use of ab initio methods to provide data points to aid in the design of the force fields makes this term somewhat inaccurate, though still widely used.

The utility of molecular mechanics depends crucially on the development of accurate force fields. Good quality force fields have been developed for a fairly broad range of compounds including many compounds of interest in biochemistry.

While we will not attempt to survey the wide range of force fields that are available, one particular subset of compounds for which good quality force fields are available involve H, C, N, O, F, Si, P, S, Cl (and perhaps a few others) when they are restricted to form chemically uncomplicated structures (e.g., bond strain is not too great, dangling bonds are few or absent, etc).

Many atomically precise structures which should be useful in nanotechnology fall in this class and can be modeled with an accuracy adequate to determine the behaviour of molecular machines. A potential energy function of particular utility in modeling diamondoid structures was described by Brenner (*Empirical potential for hydrocarbons for use in simulating the chemical vapour deposition of diamond films,* Donald W.

Brenner's potential energy function, though limited to the elements carbon and hydrogen, has the great advantage that it will handle transition states, unstable structures, and the like. Thus, given essentially arbitrary coordinates of the carbon and hydrogen atoms in a system, Brenner's potential will return the energy of the system. This permits molecular dynamical modeling of arbitrary hydrocarbon systems, including systems which use synthetic reactions involved in the synthesis of diamond.

A particular reaction of interest is the selective abstraction of a chosen hydrogen atom from a diamond surface. See *Surface patterning by atomically-controlled chemical forces: molecular dynamics simulations* by Sinnott et al., Surface Science 316 (1994), the work of Robertson et al. for an illustration of the use of Brenner's potential in modeling the behaviour of graphitic gears, including failure modes).

Brenner has commented on the utility of this potential for modeling proposed molecular machine systems. Of course, the "accuracy" of the force fields depends on the application. A force field which was accurate to (say) 10 kcal/mole would be unable to correctly predict many properties of interest in biochemistry.

For example, such a force field would lead to serious errors in predicting the correct three-dimensional structure of a protein. Given the linear sequence of amino acids in a protein, the protein folding problem is to determine how it will fold in three dimensions when put in solution. Often, the correct configuration will have an energy that differs from other (incorrect) configurations by a relatively modest amount, and so a force field of high accuracy is required. Consider, however, the bearing illustrated in figures 1 and 2. Unlike the protein folding problem, where an astronomical range of configurations of similar energy are feasible, the bearing basically has only one configuration: a bearing. While the protein has many unconstrained torsions, there are no unconstrained torsions in the bearing.

To significantly change any torsion angle in the bearing would involve ripping apart bonds. Thus, the same force field which is of marginal utility in dealing with strands of floppy protein is quite adequate for solid blocks of stiff diamondoid material.

Not only will small errors in the force field still result in an accurate prediction of the global minima, (which in this case will be a single large basin in the potential energy surface) but also the range of possible structures is so sharply limited that little or no computational effort need be spent comparing the energies of different configurations. Long computational runs to evaluate the statistical properties of ensembles of configurations are thus eliminated.

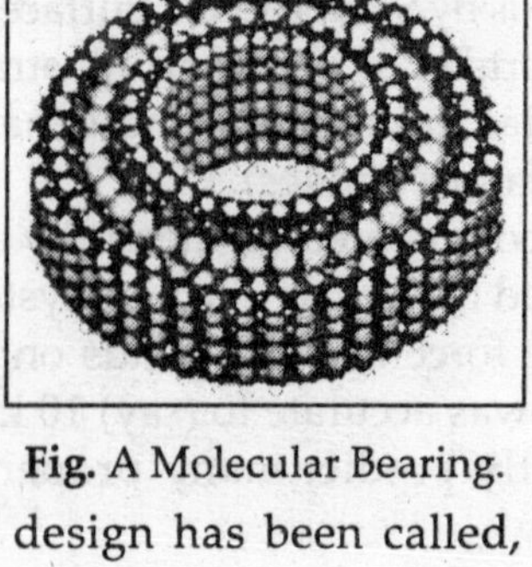

Fig. A Molecular Bearing.

This style of design has been called, only half in jest,

"molecular bridge building" because bridges are also designed with large safety margins. This observation, that the same force field that is inadequate for one class of structures is quite adequate for the design and modeling of a different class of structures, leads to a more general principle.

Computational experiments generally provide an answer with some error distribution. If the errors produced by the model are of a similar size to the errors that would result in incorrect device function, then the model is unreliable.

On the other hand, if the errors in the model are small compared with the errors that will produce incorrect device function, then the results of the model are likely to be reliable.

If a device design falls in the former category, i.e., small errors in the model will produce conflicting forecasts about device function, then the conservative course of action is to reject the proposed design and keep looking.

That is, *we can deliberately design structures which are indeed adequately handled by our computational tools.* As illustrated by the bearing, there are a wide range of basically mechanical structures whose stability is sufficiently clear and whose interactions with other structures is sufficiently simple that their behaviour can be adequately modeled with currently available force fields.

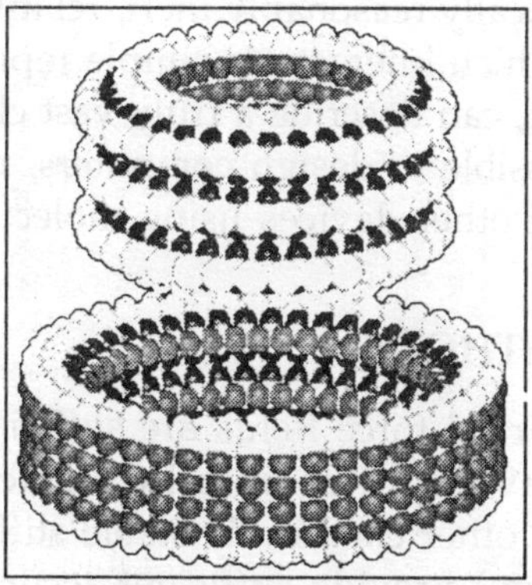

Fig. The Bearing Taken Apart.

An even stronger (although somewhat more subtle) statement is possible. The bearing illustrated in figures 1 and

2 is simply a single bearing from a very large class of bearings. The strain in the axle and the sleeve is proportional to the diameter of the bearing. By increasing the diameter, we can reduce the strain.

Thus, we can design bearings in this broad class in which the strain can be reduced to whatever level we desire. Because we are dealing with what amounts to a strained block of diamond, the fact that we can reduce strain arbitrarily means that we can design a bearing whose local structure bears as close a resemblance to an unstrained block of diamond as might be wished.

We can therefore be very confident indeed that some member of this class will perform the desired function (that of a bearing) and will work in accordance with our expectations. Given that we have developed software tools that are capable of creating and modeling most of the members of a broad class of devices, then we can investigate many individual class members.

This investigation then lets us make rather confident statements about the functionality that members of this class can provide, even if there might be residual doubts about individual class members.

A moments reflection will show that the class of objects which are chemically reasonably inert, relatively stiff (no free torsions), and which interact via simple repulsive forces that occur on contact; can describe a truly vast class of machines. Indeed, it is possible to design computers, robotic arms and a wide range of other devices using molecular parts drawn from this class.

AB INITIO METHODS

While empirical force fields are sufficiently accurate to model the behaviour of chemically stable stiff structures interacting with other chemically stable stiff structures, they do not (at present) provide sufficient accuracy to deal with chemical transitions.

Thus, if we wish to model the manufacture of a molecular part (such as the axle or sleeve of the bearing of figures 1 and

2) then we must use higher order ab initio techniques (another paper in this issue illustrates what is meant by this in detail).

These techniques impose severe constraints on the number of atoms that can be modeled (perhaps one or two dozen heavy atoms, depending on the hardware, software, and specific type of modeling being attempted), but can provide an accuracy sufficient to analyse the chemical reactions that must necessarily take place during the synthesis of large, atomically precise structures.

An analysis of the abstraction of a hydrogen atom from various structures, including isobutane (which serves as a model of the diamond surface), has been carried out to illustrate the kind of reactions that are of interest.

More generally, higher order ab initio techniques are sufficient to analyse the addition or removal of a small number of atoms from a specific site on a work piece.

Synthesis of a large object would then consist of repeated site specific applications of a small number of basic operations, where each basic operation changed the chemical structure of only a small number of atoms at a time.

Provided that we reject reaction mechanisms where the result predicted depends on errors that are smaller than can reasonably be modeled, the analysis of these basic operations can be satisfactorily carried out with current methods and hardware.

CURRENT MODELING METHODS

In summary, it is quite possible to adequately model the behaviour of molecular machines that satisfy two constraints:

- They are built from parts that are sufficiently stable that small errors in the empirical force fields do not raise significant questions about the shape or stability of the parts,
- The synthesis of the parts is done by using positionally controlled reactions, where the actual chemical reactions involve a relatively small number of atoms whose behaviour can be adequately modeled with higher order ab initio methods.

Clearly, not all molecular machines satisfy these constraints. Is the range of molecular machines which do satisfy these constraints sufficiently large to justify the effort of designing and modeling them? And in particular, can we satisfactorily model Drexler's assembler within these constraints?

The fundamental purpose of an assembler is to position atoms. To this end, it is imperative that we have models that let us determine atomic positions, and this is precisely what molecular mechanics provides. Robotic arms or other positioning devices are basically mechanical in nature, and will allow us to position molecular parts during the assembly process.

Molecular mechanics provides us with an excellent tool for modeling the behaviour of such devices. The second fundamental requirement is the ability to make and break bonds at specific sites.

While molecular mechanics provides an excellent tool for telling us where the tip of the assembler arm is located, current force fields are not adequate to model the specific chemical reactions that must then take place at the tip/work-piece interface involved in building an atomically precise part (though this statement must be modified in light of the work done with Brenner's potential, discussed above). For this, higher order ab initio calculations are sufficient.

The glaring omission in this discussion is the modeling of the kind of electronic behaviour that occurs in switching devices. Clearly, it is possible to model electronic behaviour with some degree of accuracy, and equally clearly molecular machines that are basically electronic in nature will be extremely useful.

It will therefore be desirable to extend the range of computational models discussed here to include such devices. For the moment, however, the relatively modest inclusion of electrostatic motors as a power source is probably sufficient to provide us with adequate "design space" to design and model an assembler.

While it might at first glance appear that electronics will

be required for the computational element in the assembler, in fact molecular mechanical logic elements will be sufficient. Babbage's proposal from the 1800's clearly demonstrates the feasibility of mechanical computation (it is interesting to note that a working model of Babbage's difference engine has been built and is on display at the British Museum of Science.

They were careful to use parts machined no more accurately than the parts available to Babbage, in order to demonstrate that his ideas could have been implemented in the 1800's). Drexler's analysis of a specific molecular mechanical logic element and further analysis of system design issues also make it clear that molecular mechanical computation is sufficient for the molecular computer required in an assembler.

Molecular electronic proposals for computation have not been worked out as clearly at the system level as the molecular mechanical concepts. Thus, at the moment, molecular mechanical proposals are better understood, at least in this particular context.

This situation is likely to change, and when it does the electronic designs could be incorporated (where appropriate) into the design and modeling of an assembler. However, it is not entirely obvious that electronic designs will prove superior to molecular mechanical designs, particularly when device parameters such as size and energy dissipation are considered.

While it seems virtually certain that electronic devices will prove faster than molecular mechanical devices, it is less obvious that electronic devices must necessarily be either smaller or dissipate less energy (though confer more recent work on reversible logic).

Indeed, given the difficulty of localizing individual electrons, it seems possible that electronic devices will prove to be inherently larger than molecular mechanical devices. This might provide a long term role for molecular mechanical devices as high density memory elements, or as logic elements when space (or atom count) is particularly constrained.

Whether the assembler is designed and built with an electronic or mechanical computer is less significant than designing and building it. By way of example, a stored programme computer can be built in many different ways. Vacuum tubes, transistors, moving mechanical parts, fluidics, and other methods are all entirely feasible.

Delaying the development of the Eniac because transistors are better than vacuum tubes would have been a most unwise course of action. Similarly, as we consider the design, modeling, and construction of an assembler, we should not hesitate to use simpler methods that we can understand in favour of better methods that are not yet fully in hand.

The methods of computational chemistry available today allow us to model a wide range of molecular machines with an accuracy sufficient in many cases to determine how well they will work.

We can deliberately confine ourselves to the subset of devices where our modeling tools are good enough to give us confidence that the proposals should work.

This range includes bearings, computers, robotic arms, site-specific chemical reactions utilized in the synthesis of complex structures, and more complex systems built up from these and related components. Drexler's assembler and related devices can be modeled using our current approaches and methods.

MOLECULAR COMPILERS

Computational nanotechnology includes not only the tools and techniques required to model proposed molecular machines, it must also include the tools required to specify such machines. Molecular machine proposals that would require millions or even billions of atoms have been made.

The total atom count of an assember might be roughly a billion atoms. While commerically available molecular modeling packages provide facilities to specify arbitrary structures, it is usually necessary to "point and click" for each atom involved.

This is obviously unattractive for a device as complex as an assembler with its roughly one billion atoms. It should be clear that molecular CAD tools will be required if we are to specify such complex structures in atomic detail with a reasonable amount of effort.

An essential development is that of "molecular compilers" which accept, as input, a high level description of an object and produce, as output, the atomic coordinates, atom types, and bonding structure of the object. A simple molecular compiler has already been written at PARC, and was used to produce the bearing of figures 1 and 2.

The specification of the axle was:

Scale	0.9
Tube	0 0. 75 0. 75 1. 75 0 0 0 17 – 17 0 5. 25 5. 25
Grid	0 0. 5 – 0. 5 0 1 1
Delete	1. 25 0 0
Grid	0 0. 25 0 0 0 0. 25
Change	O_3 to S_3 1. 5 0 0

The first line, "scale 0.9," is simply an instruction to shrink the size of the axle by 10% compared with the normal size. This allows the axle to be positioned inside the sleeve before joint minimization of the axle and sleeve is done (using MSI's PolyGraf, a commercially available molecular mechanics package that implements both MM2 and the Dreiding II force field).

If the axle w were not done, then the atoms in the axle and the sleeve would be comingled, and minimization would produce meaningless results. The second line begins with the "tube" specifier.

This tells the programme to produce a tubular structure, rather than a "block" or a "ring." the "ring" specifier is used to produce toroidal structures, e.g., tubes that have been further bent into donuts.

The first triplet of numbers following "tube," "0 0.75 0.75," specifies the offset to be used for the crystal lattice. The second triplet, "1.75 0 0" specifies that the surface of the tube is , and that the thickness of the tube wall is 1. 75 lattice spacings. The third triplet, "0 17 – 17," specifies the direction and length of the circumference of the tube.

The direction [0 17 – 17] (or [0 1 – 1]) is at right angles to [1 0 0], the direction of the tube surface. The circumference of the tube is simply the length of the [0 17 – 17] vector. Finally, "0 5.25 5.25" gives the direction of the axis and the length of the tube. This does not fully specify the axle of the bearing, for the 100 surface has been cut circumferentially to produce grooves. The next command, "grid 0 0.5 – 0.5 0 1 1" specifies that a grid is to be laid down on the surface.

The grid is specified by two vectors, "0 0.5 – 0.5" and "0 1 1,"" which give the directions and lengths of the edges of the unit parallelogram from which the grid is composed. The next command, "delete 1.25 0 0"" specifies that the point at coordinates "1.25 0 0" is to be deleted, and further, that all points in the same location with respect to any unit parallelogram of the grid are also to be deleted. Thus, this "grid.... delete...." cuts out the grooves that are visible on the outer surface of the axle.

Finally, the commands "grid 0 0.25 0 0 0 0.25," and "change O_3 to S_3 1.5 0 0" are used to lay down a new, very fine grid that causes all oxygen atoms on the surface of the axle to be changed to sulfur. A similar set of commands was used to specify the sleeve of the bearing.

It is easy to specify bearings with different surfaces, surface orientations, circumferences, lengths, etc. To select a (111), (110), (312), or any other surface, it is sufficient to change the vector that specifies the surface, and the vectors that specify the tangent to the surface and the axis of the tube (which must be at right angles to the surface vector).

The programme will generate output in either PolyGraf format or in Brookhaven format, so the output should be readable by most computational chemistry packages.

The software required to design and model complex molecular machines is either already available, or can be readily developed over the next few years.

Many of the modeling issues can be dealt with by existing commercially available computational chemistry packages. The molecular compilers and other molecular CAD tools needed for this work can be implemented using generally understood techniques and methods from computer science. Using this approach, it will be possible to substantially reduce the development time for complex molecular machines, including Drexler's assembler. This approach is similar in spirit to the computer aided design and modeling used to speed the development of many products today.

The author was part of a general purpose computer start up which successfully designed and built a new computer from scratch. This included the hardware, software, compilers, operating system, etc.

During this process, extensive use was made of computational models to verify each level of the design. The operating system was written in Pascal, and checked out on another computer. The compilers were written in Pascal, and also checked out on another computer.

The code produced by the compilers was checked out on an instruction set simulator. The microcode was checked out on a micro-instruction simulator. The logic design was checked out with logic level simulation tools, and circuit simulation packages were used to verify the detailed electronic design. All this work was done at the same time.

The software was written and debugged even though the machine on which it would eventually be executed didn't exist. The microcode was written and debugged before the hardware was available. When the hardware was finally made available, system integration went relatively rapidly.

Imagine, for a moment, how long it would have taken to develop this system had we carried out the development in the obvious sequential fashion. First, we would have implemented the hardware and only then begun work on the microengine.

Later, with the hardware and microengine fully checked out and working, we could have developed and debugged the microcode. With this firmly in hand, we could then have written the compilers and verified the code they produced. Finally, we could have implemented and checked out the operating system. Needless to say, such a strategy would have been very slow and tedious.

Doing things in the simple and most obvious way often takes a lot longer than is needed. If we were to approach the design and construction of an assembler using the simple serial method, it would take a great deal longer than if we systematically attacked and simultanesouly solved the problems that arise at all levels of the design at one and the same time. That is, by using methods similar to those used to design a modern computer, including intensive computational modeling of individual components and sub- systems, we can greatly shorten the time required to design and build complex molecular machines. This can be further illustrated by considering the traditional manner of growth of our synthetic capability over time (figure). Today, we find we are able to synthesize a certain range of compounds and structures.

As time goes by, we will be able to synthesize an ever larger range of structures. This growth in our ability will proceed on a broad front, and reflects the efforts of a broad range of researchers who are each pursuing individual goals without concern about the larger picture into which they might fit. As illustrated in figure, given sufficient time we will eventually be able to synthesize complex molecular machines simply because we will eventually develop the ability to synthesize just about anything.

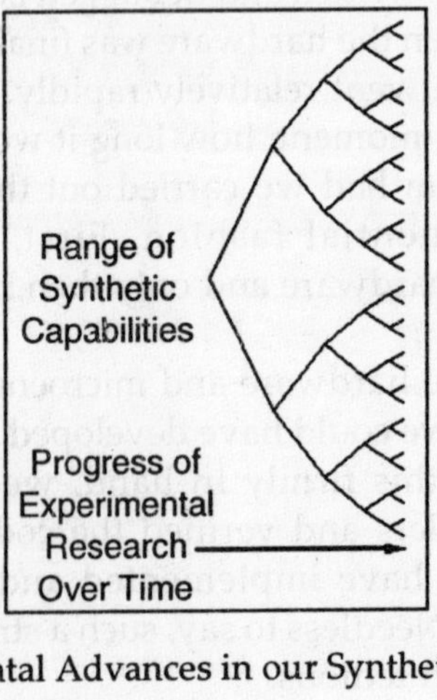

Fig. Experimental Advances in our Synthetic Capabilities

However, if we wish to develop a particular kind of device, e.g., an assembler, then we can speed the process up (as illustrated in figure) by conducting computational experiments designed to clarify the objective and to specify more precisely the path from our current range of synthetic capabilities to the objective.

Such computational experiments are inexpensive, can provide very detailed information, are possible for any structure (whether we can or cannot synthesize it) and will, in general, reduce the "time to market" for the selected product.

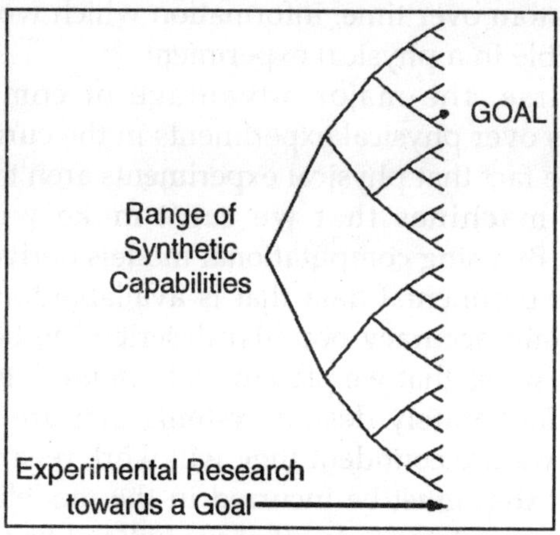

Fig. Experimental Progress Towards a Goal

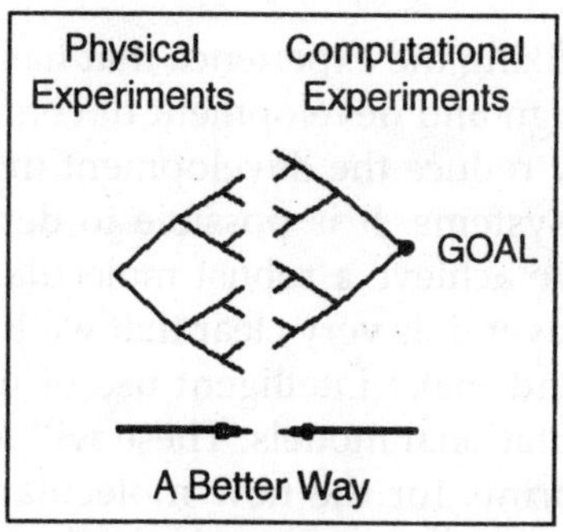

Fig. A more Efficient Method of Achieving a Goal

Computational experiments let us examine structures

quickly and easily, rejecting those which have obvious defects (a precursor to the bearing shown in figure, for example, was too strained.

By modifying the design and again minimizing the structure, we found a design with an acceptable strain). This kind of examination of the "design space" is impossible with physical experiments today, but is easily done with computational experiments. Computational experiments also provide more information. For example, molecular dynamics can literally provide information about the position of each individual atom over time, information which would usually be inaccessible in a physical experiment.

Of course, the major advantage of computational experiments over physical experiments in the current context is the simple fact that physical experiments aren't possible for molecular machines that we can't make with today's technology. By using computational models derived from the wealth of experimental data that is available today, we can (within certain accuracy bounds) describe the behaviour of proposed systems that we plan to build in the future.

If we deliberately design systems that are sufficiently robust that we are confident they will work regardless of the small errors that must be incurred in the modeling process, we can design systems today that we will not be able to build for some years, and yet still have reasonable confidence that they will work.

By fully utilizing the experience that has been developed in the rapid design and development of complex systems we can dramatically reduce the development time for molecular manufacturing systems. It is possible to debate how long it will be before we achieve a robust molecular manufacturing capability. However, it is very clear that we'll get there sooner if we develop and make intelligent use of molecular design tools and computational models. These will let us design and check the blueprints for the new molecular manufacturing technologies that we now see on the horizon, and will let us chart a more rapid and more certain path to their development.

Chapter 4

Properties of Individual Nano Particles

USING NANOSENSOR

Nanosensors are any biological, chemical, or physical sensory points used to convey information about nanoparticles to the macroscopic world. Though humans have not yet been able to synthesize nanosensors, predictions for their use mainly include various medicinal purposes and as gateways to building other nanoproducts, such as computer chips that work at the nanoscale and nanorobots.

Presently, there are several ways proposed to make nanosensors, including top-down lithography, bottom-up assembly, and molecular self-assembly.

APPLICATIONS

Medicinal uses of nanosensors mainly revolve around the potential of nanosensors to accurately identify particular cells or places in the body in need. By measuring changes in volume, concentration, displacement and velocity, gravitational, electrical, and magnetic forces, pressure, or temperature of cells in a body, nanosensors may be able to distinguish between and recognize certain cells, most notably those of cancer, at the molecular level in order to deliver medicine or monitor development to specific places in the body.

In addition, they may be able to detect macroscopic variations from outside the body and communicate these changes to other nanoproducts working within the body.

One example of nanosensors involves using the fluorescence properties of cadmium selenide quantum dots as sensors to uncover tumors within the body. By injecting a body with these quantum dots, a doctor could see where a tumor or cancer cell was by finding the injected quantum dots, an easy process because of their fluorescence.

Developed nanosensor quantum dots would be specifically constructed to find only the particular cell for which the body was at risk. A downside to the cadmium selenide dots, however, is that they are highly toxic to the body. As a result, researchers are working on developing alternate dots made out of a different, less toxic material while still retaining some of the fluorescence properties.

In particular, they have been investigating the particular benefits of zinc sulfide quantum dots which, though they are not quite as fluorescent as cadmium selenide, can be augmented with other metals including manganese and various lanthanide elements.

In addition, these newer quantum dots become more fluorescent when they bond to their target cells. Potential predicted functions may also include sensors used to detect specific DNA in order to recognize explicit genetic defects, especially for individuals at high-risk and implanted sensors that can automatically detect glucose levels for diabetic subjects more simply than current detectors.

DNA can also serve as sacrificial layer for manufacturing CMOS IC, integrating a nanodevice with sensing capabilities. Therefore, using proteomic patterns and new hybrid materials, nanobiosensors can also be used to enable components configured into a hybrid semiconductor substrate as part of the circuit assembly.

The development and miniaturization of nanobiosensors should provide interesting new opportunities. Other projected products most commonly involve using nanosensors to build smaller integrated circuits, as well as incorporating them into various other commodities made using other forms of nanotechnology for use in a variety of situations including transportation, communication, improvements in structural integrity, and robotics. Nanosensors may also eventually be valuable as more accurate monitors of material states for use in systems where size and weight are constrained, such as in satellites and other aeronautic machines.

EXISTING NANOSENSORS

Currently, the most common mass-produced functioning nanosensors exist in the biological world as natural receptors of outside stimulation. For instance, sense of smell, especially in animals in which it is particularly strong, such as dogs, functions using receptors that sense nanosized molecules.

Certain plants, too, use nanosensors to detect sunlight; various fish use nanosensors to detect minuscule vibrations in the surrounding water; and many insects detect sex pheromones using nanosensors.

Certain electromagnetic sensors have also been in use in photoelectric systems. These work because the specific sensors called, aptly, photosensors are easily influenced by light of various wavelengths.

The electromagnetic source transfers energy to the photosensors and energizes them into an excited state which causes them to release an electron into a semiconductor. At that point, it is relatively easy to detect the electricity coming from the sensors, and thus easy to know if the sensors are receiving light. Though more advanced uses of photosensors incorporating other forms of nanotechnology have yet to be implemented into consumer society, most film cameras have used photosensors at the nano size for years. Traditional film uses a layer of silver ions that become excited by solar energy and clump into groups, as small as four atoms apiece in some cases, that scatter light and appear dark on the frame.

Various other types of film can be made using a similar process to detect other specific wavelengths of light, including x-rays, infrared, and ultraviolet.

One of the first working examples of a synthetic nanosensor was built by researchers at the Georgia Institute of Technology in 1999. It involved attaching a single particle onto the end of a carbon nanotube and measuring the vibrational frequency of the nanotube both with and without the particle. The discrepancy between the two frequencies allowed the researchers to measure the mass of the attached particle.

Chemical sensors, too, have been built using nanotubes to detect various properties of gaseous molecules. Carbon nanotubes have been used to sense ionization of gaseous molecules while nanotubes made out of titanium have been employed to detect atmospheric concentrations of hydrogen at the molecular level.

Many of these involve a system by which nanosensors are built to have a specific pocket for another molecule. When that particular molecule, and only that specific molecule, fits into the nanosensor, and light is shone upon the nanosensor, it will reflect different wavelengths of light and, thus, be a different colour.

Production Methods

There are currently several hypothesized ways to produce nanosensors. Top-down lithography is the manner in which most integrated circuits are now made. It involves starting out with a larger block of some material and carving out the desired form. These carved out devices, notably put to use in specific microelectromechanical systems used as microsensors, generally only reach the micro size, but the most recent of these have begun to incorporate nanosized components.

Another way to produce nanosensors is through the bottom-up method, which involves assembling the sensors out of even more minuscule components, most likely individual atoms or molecules.

This would involve moving atoms of a particular substance one by one into particular positions which, though it has been achieved in laboratory tests using tools such as atomic force microscopes, is still a significant difficulty, especially to do en masse, both for logistic reasons as well as economic ones.

Most likely, this process would be used mainly for building starter molecules for self-assembling sensors. he third way, which promises far faster results, involves self-assembly, or "growing" particular nanostructures to be used as sensors. This most often entails one of two types of assembly.

The first involves using a piece of some previously created or naturally formed nanostructure and immersing it in free atoms of its own kind.

After a given period, the structure, having an irregular surface that would make it prone to attracting more molecules as a continuation of its current pattern, would capture some of the free atoms and continue to form more of itself to make larger components of nanosensors.

The second type of self-assembly starts with an already complete set of components that would automatically assemble themselves into a finished product. Though this has been so far successful only in assembling computer chips at the micro size, researchers hope to eventually be able to do it at the nanometer size for multiple products, including nanosensors.

Accurately being able to reproduce this effect for a desired sensor in a laboratory would imply that scientists could manufacture nanosensors much more quickly and potentially far more cheaply by letting numerous molecules assemble themselves with little or no outside influence, rather than having to manually assemble each sensor.

What are Nanorobots

Nanorobotics is emerging as a demanding field dealing with miniscule things at molecular level. Nanorobots are quintessential nanoelectromechanical systems designed to perform a specific task with precision at nanoscale dimensions.

Its advantage over conventional medicine lies on its size. Particle size has effect on serum lifetime and pattern of deposition. This allows drugs of nanosize to be used in lower concentration and has an earlier onset of therapeutic action.

It also provides materials for controlled drug delivery by directing carriers to a specific location. The typical medical nanodevice will probably be a micron-scale robot assembled from nanoscale parts.

These nanorobots can work together in response to environment stimuli and programmed principles to produce macro scale results.

ELEMENTS OF NANOROBOTS

Carbon will likely be the principal element comprising the bulk of a medical nanorobot, probably in the form of diamond or diamondoid/fullerene nanocomposites. Many other light elements such as hydrogen, sulfur, oxygen, nitrogen, fluorine, silicon, etc. will be used for special purposes in nanoscale gears and other components.

The chemical inertness of diamond is proved by several experimental studies. One such experiment conducted on mouse peritoneal macrophages cultured on DLC showed no significant excess release of lactate dehydrogenase or of the lysosomal enzyme β N-acetyl-D-glucosaminidase an enzyme known to be released from macrophages during inflammation.

Morphological examination revealed no physical damage to either fibroblasts or macrophages, and human osteoblast like cells confirming the biochemical indication that there was no toxicity and that no inflammatory reaction was elicited in vitro. The smoother and more flawless the diamond surface, the lesser is the leukocyte activity and fibrinogen adsorption. Interestingly, on the rougher "polished" surface, a small number of spread and fused macrophages were present, indicating that some activation had occurred.

The exterior surface with near-nanometer smoothness results in very low bioactivity. Due to the extremely high surface energy of the passivated diamond surface and the strong hydrophobicity of the diamond surface, the diamond exterior is almost completely chemically inert.

CONSTITUENTS AND DESIGN OF NANOROBOTS

Nanorobots will possess full panoply of autonomous subsystems whose design is derived from biological models. Drexler evidently was the first to point out, in 1981, that complex devices resemble biological models in their structural components. The various components in the nanorobot design may include onboard sensors, motors, manipulators, power supplies, and molecular computers.

Perhaps the best-known biological example of such molecular machinery is the ribosome the only freely programmable nanoscale assembler already in existence. The mechanism by which protein binds to the specific receptor site might be copied to construct the molecular robotic arm.

The manipulator arm can also be driven by a detailed sequence of control signals, just as the ribosome needs mRNA to guide its actions. These control signals are provided by external acoustic, electrical, or chemical signals that are received by the robot arm via an onboard sensor using a simple "broadcast architecture a technique which can also be used to import power. the biological cell may be regarded as an example of a broadcast architecture in which the nucleus of the cell send signals in the form of mRNA to ribosomes in order to manufacture cellular proteins.

Assemblers are molecular machine systems that could be described as systems capable of performing molecular manufacturing at the atomic scale which require control signals provided by an onboard nanocomputer This programmable nanocomputer must be able to accept stored instructions which are sequentially executed to direct the manipulator arm to place the correct moiety or nanopart in the desired position and orientation, thus giving precise control over the timing and locations of chemical reactions or assembly operations.

CONSTRUCTION OF NANOROBOTS

There are two main approaches to building at the nanometer scale: positional assembly and self-assembly. In positional assembly, investigators employ some devices such as the arm of a miniature robot or a microscopic set to pick up molecules one by one and assemble them manually.

In contrast, self-assembly is much less painstaking, because it takes advantage of the natural tendency of certain molecules to seek one another out. With self-assembling components, all that investigators have to do is put billions of them into a beaker and let their natural affinities join them automatically into the desired configurations. Making complex nanorobotic systems requires manufacturing techniques that can build a molecular structure via computational models of diamond mechanosynthesis.

DMS is the controlled addition of carbon atoms to the growth surface of a diamond crystal lattice in a vacuum-manufacturing environment. Covalent chemical bonds are formed one by one as the result of positionally constrained mechanical forces applied at the tip of a scanning probe microscope apparatus, following a programmed sequence.

Recognition of Target Site by Nanorobots

Different molecule types are distinguished by a series of chemotactic sensors whose binding sites have a different affinity for each kind of molecule. The control system must ensure a suitable performance. It can be demonstrated with a determined number of nanorobots responding as fast as possible for a specific task based scenario.

In the 3D workspace the target has surface chemicals allowing the nanorobots to detect and recognize it. Manufacturing better sensors and actuators with nanoscale sizes makes them find the source of release of the chemical. Nanorobot Control Design (NCD) simulator was developed, which is software for nanorobots in environments with fluids dominated by Brownian motion and viscous rather than inertial forces.

First, as a point of comparison, the scientists used the nanorobots' small Brownian motions to find the target by random search. In a second method, the nanorobots monitor for chemical concentration significantly above the background level. After detecting the signal, a nanorobot estimates the concentration gradient and moves toward higher concentrations until it reaches the target.

In the third approach, nanorobots at the target release another chemical, which others use as an additional guiding signal to the target. With these signal concentrations, only nanorobots passing within a few microns of the target are likely to detect the signal. Thus, we can improve the response by having the nanorobots maintain positions near the vessel wall instead of floating throughout the volume flow in the vessel from monitoring the concentration of a signal from others; a nanorobot can estimate the number of nanorobots at the target. So, the nanorobot uses this information to determine when enough nanorobots are at the target, thereby terminating any additional "attractant" signal a nanorobot may be releasing.

It is found that the nanorobots stop attracting others once enough nanorobots have responded. The amount is considered enough when the target region is densely covered by nanorobots. Thus these tiny machines work at the target site accurately and precisely to that extent only to which it is designed to do.

Nanorobots Evading the Immune System

Every medical nanorobot placed inside the human body will encounter phagocytic cells many times during its mission. Thus all Nanorobots, which are of a size capable of ingestion by phagocytic cells, must incorporate physical mechanisms and operational protocols for avoiding and escaping from phagocytes. The initial strategy for medical nanorobots is first to avoid phagocytic contact or recognition.

To avoid being attacked by the host's immune system, the best choice is to have an exterior coating of passive diamond. The smoother and flawless the coating, the lesser is the reaction from the body's immune system. And if this fails then to avoid it's binding to the phagocyte surface that leads to phagocytic activation.

If trapped, the medical nanorobot can induce exocytosis of the phagosomal vacuole in which it is lodged or inhibit both phagolysosomal fusion and phagosome metabolism. In rare circumstances, it may be necessary to kill the phagocyte or to blockade the entire phagocytic system.

The most direct approach for a fully functional medical nanorobot is to employ its motility mechanisms to locomote out of, or away from, the phagocytic cell that is attempting to engulf it. This may involve reverse cytopenetration, which must be done cautiously (e.g., the rapid exit of nonenveloped viruses from cells can be cytotoxic). It is possible that frustrated phagocytosis may induce a localized compensatory granulomatous reaction.

Medical nanorobots therefore may also need to employ simple but active defensive strategies to forestall granuloma formation. Metabolizing local glucose and oxygen for energy can do the powering of the nanorobots. In a clinical environment, another option would be externally supplied acoustic energy.

When the task of the nanorobots is completed, they can be retrieved by allowing them to exfuse themselves via the usual human excretory channels or can also be removed by active scavenger systems.

Nanorobots in Cancer Detection and Treatment

The development of nanorobots may provide remarkable advances for diagnosis and treatment of cancer. Nanorobots could be a very helpful and hopeful for the therapy of patients, since current treatments like radiation therapy and chemotherapy often end up destroying more healthy cells than cancerous ones

From this point of view, it provides a non-depressed therapy for cancer patients. The Nanorobots will be able to distinguish between different cell types that is the malignant and the normal cells by checking their surface antigens (they are different for each type of cell).

This is accomplished by the use of chemotactic sensors keyed to the specific antigens on the target cells. Another approach uses the innovative methodology to achieve decentralized control for a distributed collective action in the combat of cancer.

Using chemical sensors they can be programmed to detect different levels of E-cadherin and β -catenin in primary and metastatic phases. Medical nanorobots will then destroy these cells, and only these cells.

The following control methods were considered:

- *Random*: Nanorobots moving passively with the fluid

reaching the target only if they bump into it due to Brownian motion.

- *Follow gradient*: Nanorobots monitor concentration intensity for E-cadherin signals, when detected, measure and follow the gradient until reaching the target. If the gradient estimate subsequent to signal detection finds no additional signal in50ms, the nanorobot considers the signal to be a false positive and continues flowing with the fluid.
- Follow gradient with attractant: as above, but nanorobots arriving at the target, they release in addition a different chemical signal used by others to improve their ability to find the target. Thus, a higher gradient of signal intensity of E-cadherin is used as chemical parameter identification in guiding nanorobots to identify malignant tissues. Integrated nanosensors can be utilized for such a task in order to find intensity of E-cadherin signals. Thus they can be employed effectively for treating cancer.

Cancer Detection and Treatment

Pharmacyte is a self-powered, computer controlled medical nanorobot system capable of digitally precise transport, timing, and targeted-delivery of pharmaceutical agents to specific cellular and intracellular destinations within the human body.

Pharmacytes escape the phagocytic process as they will not embolize small blood vessels because the minimum viable human capillary that allows passage of intact erythrocytes and white cells is 3–4 micronmeter in diameter, which is larger than the largest proposed Pharmacyte.

Pharmacytes will have many applications in nano-medicine such as initiation of apoptosis in cancer cells and direct control of cell signaling processes. Pharmacytes could also tag target cells with biochemical natural defensive or scavenging systems, a strategy called "phagocytic flagging".

For example, novel recognition molecules are expressed on the surface of apoptotic cells. In the case of T lymphocytes, one such molecule is phosphatidylserine, a lipid that is normally restricted to the inner side of the plasma membrane [1m] but, after the induction of apoptosis, appears on the outside. Cells bearing this molecule on their surface can then be recognized and removed by phagocytic cells.

Seeding the outer wall of a target cell with phosphatidylserine or other molecules with similar action could activate phagocytic behaviour by macrophages, which had mistakenly identified the target cell as apoptotic substances capable of triggering a reaction by the body Pharmacytes would be capable of carrying up to approximately 1cubicmeter of pharmaceutical payload stored in onboard tanks that are mechanically offloaded using molecular sorting pumps operated under the control of an onboard computer.

Depending on mission requirements, the payload can be discharged into the proximate extracellular fluid or delivered directly into the cytosol using a transmembrane injector mechanism. If needed for a particular application, deployable mechanical cilia and other locomotive systems can be added to the Pharmacyte to permit transvascular and transcellular mobility, thus allowing delivery of pharmaceutical molecules to specific cellular and even intracellular addresses with negligible error.

Pharmacytes, once depleted of their payloads or having completed their mission, would be recovered from the patient by conventional excretory pathways. The nanorobots might then be recharged, reprogrammed and recycled for use in a second patient who may need a different pharmaceutical agent targeted to different tissues or cells than in the first patient.

Nanorobots in the Diagnosis and Treatment of Diabetes

Glucose carried through the blood stream is important to maintain the human metabolism working healthfully, and its correct level is a key issue in the diagnosis and treatment of diabetes. Intrinsically related to the glucose molecules, the protein hSGLT3 has an important influence in maintaining proper gastrointestinal cholinergic nerve and skeletal muscle function activities, regulating extracellular glucose concentration. The hSGLT3 molecule can serve to define the glucose levels for diabetes patients.

The most interesting aspect of this protein is the fact that it serves as a sensor to identify glucose. The simulated nanorobot prototype model has embedded Complementary Metal Oxide semi-conductor (CMOS) nanobioelectronics.

It features a size of ~2 micronmeter, which permits it to operate freely inside the body. Whether the nanorobot is invisible or visible for the immune reactions, it has no interference for detecting glucose levels in blood stream. Even with the immune system reaction inside the body, the nanorobot is not attacked by the white blood cells due biocompatibility.

For the glucose monitoring the nanorobot uses embedded chemosensor that involves the modulation of hSGLT3 protein glucosensor activity. Through its onboard chemical sensor, the nanorobot can thus effectively determine if the patient needs to inject insulin or take any further action, such as any medication clinically prescribed.

The image of the NCD simulator workspace shows the inside view of a venule blood vessel with grid texture, red blood cells (RBCs) and nanorobots. They flow with the RBCs through the bloodstream detecting the glucose levels.

At a typical glucose concentration, the nanorobots try to keep the glucose levels ranging around 130 mg/dl as a target for the Blood Glucose Levels (BGLs). A variation of 30mg/dl was adopted as a displacement range, though this can be changed based on medical prescriptions.

In the medical nanorobot architecture, the significant measured data can be then transferred automatically through the RF signals to the mobile phone carried by the patient. At any time, if the glucose achieves critical levels, the nanorobot emits an alarm through the mobile phone.

Controlling Glucose Level using Nanorobots

In the simulation, the nanorobot is programmed also to emit a signal based on specified lunch times, and to measure the glucose levels in desired intervals of time.

The nanorobot can be programmed to activate sensors and measure regularly the BGLs early in the morning, before the expected breakfast time. Levels are measured again each 2 hours after the planned lunchtime. The same procedures can be programmed for other meals through the day times.

A multiplicity of blood borne nanorobots will allow glucose monitoring not just at a single site but also in many different locations simultaneously throughout the body, thus permitting the physician to assemble a whole-body map of serum glucose concentrations. Examination of time series data from many locations allows precise measurement of the rate of change of glucose concentration in the blood that is passing through specific organs, tissues, capillary beds, and specific vessels.

This will have diagnostic utility in detecting anomalous glucose uptake rates which may assist in determining which tissues may have suffered diabetes-related damage, and to what extent. Other onboard sensors can measure and report diagnostically relevant observations such as patient blood pressure, early signs of tissue gangrene, or changes in local metabolism that might be associated with early-stage cancer. Whole-body time series data collected during various patient activities levels could have additional diagnostic value in assessing the course and extent of disease.

This important data may help doctors and specialists to supervise and improve the patient medication and daily diet. This process using nanorobots may be more convenient and safe for making feasible an automatic system for data collection and patient monitoring. It may also avoid eventually infections due the daily small cuts to collect blood samples, possibly loss of data, and even avoid patients in a busy week to forget doing some of their glucose sampling. These Recent developments on nanobioelectronics show how to integrate system devices and cellular phones to achieve a better control of glucose levels for patients with diabetes.

Respirocyte - An Artificial Oxygen Carrier Nanorobot

The artificial mechanical red cell, "Respirocyte" is an

imaginary nanorobot, floats along in the blood stream. These atoms are mostly carbon atoms arranged as diamond in a porous lattice structure inside the spherical shell. The Respirocyte is essentially a tiny pressure tank that can be pumped full of oxygen (O_2) and carbon dioxide (CO_2) molecules.

Later on, these gases can be released from the tiny tank in a controlled manner. The gases are stored onboard at pressures up to about 1000 atmospheres. Respirocyte can be rendered completely nonflammable by constructing the device internally of sapphire, a flameproof material with chemical and mechanical properties otherwise similar to diamond.

There are also gas concentration sensors on the outside of each device. When the nanorobot passes through the lung capillaries, O_2 partial pressure is high and CO_2 partial pressure is low, so the onboard computer tells the sorting rotors to load the tanks with oxygen and to dump the CO_2. When the device later finds itself in the oxygen-starved peripheral tissues, the sensor readings are reversed.

That is, CO_2 partial pressure is relatively high and O_2 partial pressure relatively low, so the onboard computer commands the sorting rotors to release O_2 and to absorb CO_2.Respirocytes mimic the action of the natural hemoglobin-filled red blood cells. But a Respirocyte can deliver 236 times more oxygen per unit volume than a natural red cell. This nanorobot is far more efficient than biology, mainly because its diamondoid construction permits a much higher operating pressure. So the injection of a 5 cm^3 dose of 50% Respirocyte aqueous suspension into the bloodstream can exactly replace the entire O_2 and CO_2 carrying capacity of the patient's entire 5,400 cm^3 of blood.

Respirocyte will have pressure sensors to receive acoustic signals from the doctor, who will use an ultrasound-like transmitter device to give the Respirocyte commands to modify their behaviour while they are still inside the patient's body.

Artificial Phagocytes

A microbivore has been described, whose primary

function is to destroy microbiological pathogens found in the human bloodstream, using the "digest and discharge" protocol. Nanorobotic artificial hypothetical phagocytes called "microbivores" could patrol the bloodstream, seeking out and digesting unwanted pathogens including bacteria, viruses, or fungi. Microbivores when given intravenously (I.V) would achieve complete clearance of even the most severe septicemic infections in hours or less.

This is far better than the weeks or months needed for antibiotic-assisted natural phagocytic defenses. The nanorobots do not increase the risk of sepsis or septic shock because the pathogens are completely digested into harmless simple sugars, monoresidue amino acids, mononucleotides, free fatty acids and glycerol, which are the biologically inactive effluents from the nanorobot.

A Hypothetical Mobile Cell-Repair Nanorobot

Another nanorobot, the Chromallocyte would replace entire chromosomes in individual cells thus reversing the effects of genetic disease and other accumulated damage to our genes, preventing aging.

Chromallocyte is a hypothetical mobile cell-repair nanorobot capable of limited vascular surface travel into the capillary bed of the targeted tissue or organ, followed by extravasation, histonatation, cytopenetration, and complete chromatin replacement in the nucleus of one target cell, and ending with a return to the bloodstream and subsequent extraction of the device from the body, completing the cell repair mission."

Inside a cell, a repair machine will first size up the situation by examining the cell's contents and activity, and then take action. By working along molecule-by-molecule and structure-by-structure, repair machines will be able to repair whole cells. By working along cell-by-cell and tissue-by-tissue, they (aided by larger devices, where need be) will be able to repair whole organs.

By working through a person organ by organ, they will restore health. Because molecular machines will be able to build molecules and cells from scratch, they will be able to repair even cells damaged to the point of complete inactivity.

Further Applications of Nanorobots

Nanorobots could be used to maintain tissue oxygenation in the absence of respiration, repair and recondition the human vascular tree eliminating heart disease and stroke damage, perform complex nanosurgery on individual cells, and instantly st nch bleeding after traumatic injury.

Monitoring nutrient concentrations in the human body is a possible application of nanorobots in medicine. One of interesting nanorobot utilization is also to assist inflammatory cells (or white cells) in leaving blood vessels to repair injured tissues. Nanorobots might be used as well to seek and break kidney stones. Nanorobots could also be used to process specific chemical reactions in the human body as ancillary devices for injured organs. Nanorobots equipped with nanosensors could be developed to deliver anti-HIV drugs. Another important capability of medical nanorobots will be the ability to locate stenosed blood vessels, particularly in the coronary circulation, and treat them mechanically, chemically, or pharmacologically.

To cure skin diseases, a cream containing nanorobots may be used. It could remove the right amount of dead skin, remove excess oils, add missing oils, apply the right amounts of natural moisturizing compounds, and even achieve the elusive goal of 'deep pore cleaning' by actually reaching down into pores and cleaning them out.

The cream could be a smart material with smooth-on, peel-off convenience. A mouthwash full of smart nanomachines could identify and destroy pathogenic bacteria while allowing the harmless flora of the mouth to flourish in a healthy ecosystem. Further, the devices would identify particles of food, plaque, or tartar, and lift them from teeth to be rinsed away. Being suspended in liquid and able to swim about, devices would be able to reach surfaces beyond reach of toothbrush bristles or the fibers of floss.

As short-lifetime medical nanodevices, they could be built

to last only a few minutes in the body before falling apart into materials of the sort found in foods. Medical nanodevices could augment the immune system by finding and disabling unwanted bacteria and viruses.

When an invader is identified, it can be punctured, letting its contents spill out and ending its effectiveness. If the contents were known to be hazardous by themselves, then the immune machine could hold on to it long enough to dismantle it more completely. Devices working in the bloodstream could nibble away at arteriosclerotic deposits, widening the affected blood vessels. Cell herding devices could restore artery walls and artery linings to health, by ensuring that the right cells and supporting structures are in the right places.

This would prevent most heart attacks. Nanorobots could be used in precision treatment and cell targeted delivery, in performing nanosurgery, and in treatments for hypoxemia and respiratory illness, dentistry, bacteremic infections, physical trauma, gene therapy via chromosome replacement therapy and even biological aging. It has been suggested that a fleet of nanorobots might serve as antibodies or antiviral agents in patients with compromised immune systems, or in diseases that do not respond to more conventional measures.

There are numerous other potential medical applications, including repair of damaged tissue, unblocking of arteries affected by plaques, and perhaps the construction of complete replacement body organs.

Nanoscale systems can also operate much faster than their larger counterparts because displacements are smaller; this allows mechanical and electrical events to occur in less time at a given speed. Nanotechnology as a diagnostic and treatment tool for patients with cancer and diabetes showed how actual developments in new manufacturing technologies are enabling innovative works which may help in constructing and employing nanorobots most effectively for biomedical problems. Nanorobots applied to medicine hold a wealth of promise from eradicating disease to reversing the aging process (wrinkles, loss of bone mass and age-related conditions are all treatable at the cellular level); nanorobots are also candidates for industrial applications. The advent of molecular nanotechnology will again expand enormously the effectiveness, comfort and speed of future medical treatments while at the same time significantly reducing their risk, cost, and invasiveness.

CELLULAR BIOSCANNING

The goal of cellular bioscanning is the noninvasive and non-destructive in vivo examination of interior biological structures. One of the most common nanomedical sensor tasks is the scanning of cellular and subcellular structures.

Such tasks may include localization and examination of cytoplasmic and nuclear membranes, as well as the identification and diagnostic measurement of cellular contents including organelles and other natural molecular devices, cytoskeletal structures, biochemical composition and the kinetics of the cytoplasm.

The precision and speed of medical nanodevices is so great that they can provide a surfeit of detailed diagnostic information well beyond that which is normally needed in classical medicine for a complete analysis of somatic status.

Except in the most subtle cases a malfunction in any one component of the cellular machinery normally provokes a cascade of pathological observables in many other subsystems. Detection of any one of these cascade observables, if sufficiently unique and well-defined, may provide adequate diagnostic information to plan a proper reparative procedure.

DNA is one of the few cellular components that is regularly inspected and repaired. Most homeostatic systems adopt a more simple philosophy of periodic replacement of components regardless of functionality.

Nanomedicine allows the philosophy of inspection and repair to be extended to all cellular components. The following discussion briefly describes a few of the sensory techniques that may be useful in achieving this objective.

CELLULAR TOPOGRAPHICS

Tactile topographic scanning provides the most direct

means for examining cellular structures in vivo. For example, ex vivo live cell scanning in air by commercially available atomic force microscopy using a 20 – 40 nm radius tip allows nondestructive feature resolutions of ~50 nm across the top 10 nm of the cell.

Investigators have used the scanning tip to punch a hole in the cell, then pull back and scan the breach, observing the membrane heal itself via self-assembly in real time. Up to ~KHz scanning frequencies are possible, with up to 1024 data points per scan line.

Assuming a scan rate of ~10^6 pixels/sec, a micron-scale nanodevice, once securely anchored to the surface of a (~20 micron)^3 human cell, could employ a tactile scanning probe to image the 0.1% of plasma membrane lying within its (1.4 micron)^2 vicinity in ~2 sec to ~1 nm^2 resolution (~1 mm/sec tip velocity), or ~50 sec to ~0.2 nm (e.g., atomic) resolution (~0.2 mm/sec tip velocity).

Inside the cell, and again post-anchoring, an entire 6 micron^2 mitochondrial surface could be imaged to atomic resolution in ~100 sec; the surface of a 100-nm length of 25-nm diameter microtubule could be atomically resolved in 0.2 sec — though of course in a living cell these structures may be changing dynamically during the scanning process.

From Eqn., continuous power dissipation of a (0.1 micron)^2 scan head moving at 1 mm/sec through water is ~0.002 pW (~kT/pixel at ~10^6 pixels/sec), though of course the energy cost of sensing, recording, and processing each pixel must be at least ~10 kT/pixel (ignoring the possibility of pre-computational image compression), so the total scanner power draw could be as high as 0.02 – 0.1 pW in continuous operation. Special scanning tips and techniques should allow topographic, roughness, elastic, adhesive, chemical, electrostatic, conductance, capacitance, magnetic, or thermal surface properties to be measured.

For large (~0.1 – 1 micron) cellular components, identification and preliminary diagnosis of improper structure may be possible using measured surface characteristics which may be matched to entries in an extensive onboard library, perhaps combined with dynamic monitoring of anomalies in cross-membrane molecular traffic using chemical nanosensors.

Most membranes are self-sealing, so it should also be possible to gently insert a telescoping member into the target organelle, which member then slowly reticulates and extrudes smaller probes with sensory tips in a fixed pattern and step size, allowing the acquisition of detailed internal structural and compositional information.

Small (~1 – 10 nm) protein-based components of the cell such as enzymes, MHC carriers, and ribosomes are self-assembling or require only modest assistance to self-assemble.

Reversible mechanical denaturation of these smaller proteins, using diamondoid probe structure to displace water molecules and by using specialized handling tools akin to functionalized AFM tips and molecular clamps, may be followed by precise nondestructive amino acid sequencing to allow identification and diagnostic compositional analysis.

Afterward, the protein molecule may be refolded back into its original minimum-energy conformation, possibly with the assistance of chaperone-like structures in some cases.

Transcellular Acoustic Microscopy

Acoustic microscopy is another noninvasive scanning technique that can provide nanomedically useful spatial resolutions. Frequencies are in the GHz range, far exceeding the relaxation time of the protoplasm.

A cryogenic acoustic microscope operated at 8 GHz has demonstrated 20 nm lateral resolution in liquid helium. By 1998, the best resolution achieved with water as the coupling fluid has been 240 nm at a frequency of 4.4 GHz.

Operated in water at 310 K, a nanomechanical 1.5 GHz acoustic microscope would achieve a far-field minimum lateral resolution l/2 ~ 500 nm, ~10^5 voxels per human cell, sufficient to locate and count all major organelles and a few intermediate-scale structures.

Scanning acoustic microscopes (SAM) operated at 2 GHz in reflection mode (which allows detection of interference effects) achieve 30 – 50 nm resolution in the direction of the acoustical axis, and the Subtraction SAM approach reveals topographical deviations of 7.5 nm at 1 GHz. It has been proposed that picosecond ultrasonics could be used to obtain an image of the cytoskeleton with detail comparable to that of conventional X-ray images of a human skeleton. Power requirements are a significant constraint on acoustic reflection microscopy (echolocation).

In the simplest case, consider an acoustic emitter of radius r_E which converts an input power of P_{in} into acoustic power of intensity $I_E \leq I_{max} = 1000$ watts/m^2 with efficiency e%, conservatively taken here to be 0.50 (50%). The emitter produces a train of omnidirectional pressure pulses of amplitude $A_p = (2 \rho v_{sound} I_E)^{1/2}$ (N/m^2) and frequency ν which travels a distance X_{path} to a target of radius r_T.

As the signal reaches the target, its amplitude has been reduced by (r_E / X_{path}) due to the $1/r^2$ dependence of intensity in spherical waves and by $e^{-\alpha tiss\ v\ Xpath}$ by attenuation with $\alpha_{tiss} = 8.3 \times 10^{-6}$ sec/m for soft tissue.

Upon reaching the target, a fraction $f_{reflect}$ of the signal is reflected as a point source echo; if the target surface has an acoustic impedance similar to liver tissue and the medium is similar to water, $f_{reflect} \sim 0.05$ (5%). The echo then travels a distance X_{path} back to a receiver (which may or may not be located near the emitter), the amplitude again losing (r_T / X_{path}) by geometry and $e^{-\alpha tiss\ v\ Xpath}$ by attenuation.The echo is finally detected by the receiver which can measure a pulse of minimum pressure amplitude $DP_{min} \sim 10^{-6}$ atm.

Combining these relations gives the following result:

$$P_{in} \neq \frac{\pi \Delta P_{min}^2 X_{path}^4}{2\rho\nu_{second} e\% f_{reflece}^2 r_T^2 e^{-4} \alpha_{max}\ \nu x_{path}}$$

(watts)

To scan the entire interior of a (20 micron)^3 cell, whether from within or without, requires $X_{path} \geq 20$ microns. A frequency of n = 1.5 GHz allows $v_{sound}/v \sim 1$ micron spatial resolution. The fastest possible pulse repetition time is $X_{path}/v_{sound} \sim 13$ nanosec.

Assuming $r_E = r_T = 0.5$ micron, $P_{in} \sim 7$ pW or ~5 watts/m^2, well within the safe intensity range for acoustic radiation. At maximum safe transmitter intensity I_{max}, the longest scannable path length at 1.5 GHz is X_{path} ~ 46 microns. Power constraints are somewhat less severe in the case of acoustic transmission microscopy, a technique which requires a minimum of two physically separated components.

Consider an acoustic emitter radiating a series of omnidirectional pressure pulses of frequency v which travel across a tissue cell of width X_{path}, to be detected by a receiver on the other side after a signal transit time $t_{cell} \sim X_{path}/v_{cell}$, where v_{cell} is the speed of sound in the cytosol.

In the simplest case, assume that the cytosol is clear except for one target of interest along the path in which the speed of sound is v_{target} # v_{cell}. The minimum detectable difference in travel times between a signal that passes through the target and one that does not is Dt (~10^{-9} sec for diamondoid systems), so the minimum resolvable target size is $r_{min} \sim v_{target}\, v_{cell}\, \Delta t$ / 2 abs($v_{target} - v_{cell}$) when $r_{min} \geq v_{cell}/v$.

For soft tissue targets in water at 1.5 GHz, r_{min} ~ 30 microns; for tooth enamel in water, r_{min} ~ 1 micron. However, the use of sampling gates will permit the detection of phase shifts that are much less than the characteristic time constant of the system in incoming waves, hence the minimum thickness of objects in theory detectable by acoustic tomography can be far smaller than the values for r_{min} estimated above.

Computing acoustic tomographic power requirement P_{in} in a similar manner as for echolocation, except that the transmitted signal rather than the echo is detected and assuming the signals are of approximately equal strength regardless of the path taken, gives:

$$P_{in} \neq \frac{\pi \Delta P_{min}^2 X_{path}^4}{2\rho\, \nu_{cell} e\% r_r^2 e^{-2\alpha_{max} v x_{path}}}$$

(watts)

Using the same values as the previous example, P_{in} = 1 pW for X_{path} ~ 45 microns; for X_{path} ~ 140 microns, P_{in} = 1000 pW, near I_{max}.

In either mode of operation, data gathering is accelerated

by positioning more than one receiver around the target. Sensitivity is boosted by increasing emitter size, taking multiple measurements, or by illuminating the target with some large external source located elsewhere in the tissue or even outside of the organ.

However, cells and body tissues are mesoscopic "junkyards" — highly heterogeneous media which may produce large numbers of nontarget scattering events, thus increasing the difficulty of extracting signal from noise.

Additional complications arise due to:

- Scattering on rough surfaces;
- Rapid pressure variations with range in the Fresnel zone or near field of the transmitted signal;
- Cytoplasmic viscosity inhomogeneities due to the asymmetric arrangement of cytoskeletal structure, granules, vacuoles, and the endomembrane system; and
- Variation in cytoplasmic Young's modulus due to time-varying tensions in semirandomly-distributed cytosolic fibrillar elements which alter elasticity and thus the local speed of sound.

Magnetic Resonance Cytotomography

Electrostatic scanning is largely ineffective because of Debye-Huckel shielding. Magnetic stray field probes allow resolution of 10 nm physical features, but only for materials with substantial magnetic domains — which most biological substances lack.

But nuclear magnetic resonance (NMR) imaging may allow cellular tomography by creating 3-D proton (hydrogen atom) density maps. Atomic density maps of other biologically important elements with nonzero nuclear magnetic moments (including D, Li^6, B^{10}, B^{11}, C^{13}, N^{14}, N^{15}, O^{17}, F^{19}, Na^{23}, Mg^{25}, P^{31}, Cl^{35}, K^{39}, Fe^{57}, Cu^{63} and Cu^{65}) may also be compiled.

For example, sodium imaging is already used clinically to assess brain damage in patients with strokes, epilepsy, and tumors. In a hypothetical NMR cytotomographic nanoinstrument, a large permanent magnet is positioned near the surface of the cell or organelle to be examined.

This creates a large static background magnetic field that polarizes the protons. The spatial gradient of this field establishes a unique resonant frequency, called the Larmor frequency, within each isomagnetic surface throughout the test volume.

A second time-varying magnetic driver field is then scanned through the full range of resonant frequencies, exciting into resonance the protons in each isomagnetic surface, in turn. Depending on the sensor implementation chosen, each resonance detected may cause absorption of driver field energy, an increase in measured impedance, or even a return echo of magnetic energy as the excited protons relax to equilibrium in ~1 sec.

The large polarizing magnet is then rotated to a new orientation, moving the isomagnetic surfaces to new positions within the test volume, and the scan is repeated. A 3-D map of the spatial proton distribution may be computed after several scan cycles.

Only those regions within a linewidth of resonance DB will generate an appreciable signal, and the expected spatial resolution (say, along the z-axis) Dz = DB /(dB/dz), where dB/dz is the spatial flux gradient of B. For a polarizing field B = 1.4 tesla placed adjacent to a (20 micron)^3 cell, the cross-cell gradient dB/dz ~ 7 × 10^4 tesla/m; placed next to a (1 micron)^3 organelle, dB/dz ~ 1.4 × 10^6 tesla/m.

Assuming an effective NMR linewidth ~ 0.002 tesla, minimum spatial resolutions are 29 nm and 1.4 nm, respectively. Unfortunately, the smallest reliably detectable energy in the sensor element is ~ kT e^{SNR}. The energy required to flip a single proton is $E_{flip} = 4 \pi \nu_L L_{proton}$, where L_{proton} = 5.28 × 10^{35} joule-sec (the quantized spin angular momentum of the proton) and the Larmor resonance frequency $\nu_L = g_{proton}$ B; g_{proton} is the gyromagnetic ratio and B is the polarizing magnetic flux.

However, the population difference between spin-up and spin-down nuclei in NMR is very small. For small energy differences, the Boltzmann distribution only allows a fraction

E_{flip} /2 kT to be flipped before the upper and lower state populations are made equal and the absorption disappears. In order to flip $N_{flip} \sim kT\, e^{SNR} / 4\pi\, v_L\, L_{proton}$ protons, the sample volume must include at least:

$$N_{min} - 2e^{SNR}\left(\frac{kT}{4\pi\nu_L L_{proton}}\right)^2$$

(protons)

For T = 310 K, SNR = 2 and B = 1.4 tesla, v_L = 60 MHz and $N_{min} \sim 1.7 \times 10^{11}$ protons.

When scanning a cell or organelle, the vast majority of the protons present are in water molecules. Water at 310 K has a proton density $n_{water} = 6.7 \times 10^{28}$ protons/m^3, compared to $n_{fat} \sim 6.4 \times 10^{28}$ protons/m^3 (palmitic acid). Distinguishing fat from water thus requires a minimum sample volume of $N_{min}/(n_{water} - n_{fat}) \sim 55$ micron^3.

Proteins range from 4.4×10^{28} protons/m^3 (aspartic acid minus one water) to 7.6×10^{28} protons/m^3 (leucine minus one water), average $n_{protein} \sim 5.6 \times 10^{28}$ protons/m^3; $n_{carbohydrate} \sim 5.8 \times 10^{28}$ protons/m^3 (glucose minus one water), allowing NMR cellular tomography to resolve minimum feature sizes of 2-4 microns (560 micron^3).

This resolution may include intracellular structures such as the endoplasmic reticulum (rough ER ~1100 micron^3, smooth ER ~400 micron^3), the Golgi complex (~500 micron^3), the nucleus (~270 micron^3), possibly the nucleolus (~50 micron^3), and even large localized ferritin granule concentrations.

In 1998, the minimum sample size for microcoil (~470 micron diameter) NMR scanning was ~10,000 micron^3 or ~10^{16} protons for a 1-minute measurement cycle. Individual T-cell vacuoles ~1 micron in diameter labeled with dextran-coated iron oxide particles have

Near-Field Optical Nanoimaging

Electromagnetic waves of optical wavelength l that interact with an object are diffracted into two components, called "far-field" and "near-field." The propagation of electromagnetic radiation over distances z > l acts as a spatial filter of finite bandwidth, resulting in the familiar diffraction-limited resolution ~λ/2.

Classical optics is concerned with this far-field regime with low spatial frequencies < 2/l, and conventional optical imaging will be difficult at the cellular level in vivo. (Short-wave X-rays will damage living biological cells.)

Information about the high spatial frequency components of the diffracted waves is lost in the far-field regime, so information about sub-wavelength features of the object cannot be retrieved in classical microscopy.

However, for propagation over distances $z << \lambda$, far higher spatial frequencies can be detected because their amplitudes are then of the same order as the sample ($z = 0$). This second diffraction component is the "near-field" evanescent waves with high spatial frequencies >2/λ.

Evanescent waves are confined to subwavelength distances from the object. Thus a localized optical probe, such as a subwavelength aperture in an opaque screen, can be scanned raster fashion in this regime to generate an image with a resolution on the order of the probe size.

The original Near-field Scanning Optical Microscope (NSOM) surpassed the classical diffraction limit by operating an optical probe at close proximity to the object. The NSOM probe uses an aluminum-coated "light funnel" scanned over the sample. Visible light emanates from the narrow end (~20 nm in diameter) of the light funnel and either reflects off the sample or travels through the sample into a detector, producing a visible light image of the surface with ~12 nm resolution at $\lambda = 514.5$ nm provided the distance between light source and sample is very short, about 5 nm, with signal intensities up to 10^{11} photons/sec (~50 nanowatts).

This represents a resolution of ~λ/40. The near-field acoustic equivalent is found in the medical stethoscope, which exhibits a resolution of ~λ/100.Applications include dynamical studies at video scanning rates, low-noise high-resolution spectroscopy, and differential absorption measurements.

Optical imaging of individual dye molecules has already been demonstrated, with the ability to determine the orientation and depth of each target molecule located within ~30 nm of the scanned surface.

Molecules with nanometer-scale packing densities have been resolved to ~0.4-nm diameters using STMs to create photon emission maps, and laser interferometric NSOMs have produced clear optical images of dispersed oil drops on mica to ~1 nm resolution.

Live specimen 250-nm optical sectioning for three-dimensional dynamic imaging has also been demonstrated. Submicron laser emitters have been available since the late 1980s. It should be possible to use NSOM-like nanoprobes to optically scan the surfaces of cells or organelles to <~1 nm resolution, mapping their topography and spectroscopic characteristics to depths of tens of nanometers without penetrating the surface.

However, a proper membrane-sealing invasive light funnel ~20 nm in diameter might not seriously disrupt some cellular or organelle membranes and thus could be inserted into the interiors of these bodies or through the cytoskeletal interstices to permit deeper volumetric scanning.

The thermal conductivity of water at 310 K is 0.623 watts/m-K and the energy per 500-nm photon is 400 zJ, so for 1 micron^3 of watery tissue the maximum scan rate is ~35 micron^3/sec-K, if e^{SNR} photons are used to image each 1 nm^3 voxel with SNR = 2.

Thus a 1-sec volumetric optical scan of an aqueous ~1-micron^3 sample volume to 1 nm^3 resolution requires a ~1 nanowatt scanner running at ~GHz bit rates, raising sample volume temperature by ~0.03 K.

NSOM permits the determination of five of the six degrees of freedom for each molecule, lacking only the optically inactive rotation around the dipole axis. In principle, it should be possible to produce <~1 nm resolution near-field optical scans of in situ protein molecules, since with atomically precise fabrication and single lines of atoms as conductors a minimum light guide (metal-dielectric-metal) is 3 atoms wide.

Given a molecular laser and adequate collimation, photons passing through folded proteins will scatter according to the molecular structure. Detection of sufficient photons comprising these scattering patterns should allow the noninvasive determination of protein structure; polarized photons provide information on chirality.

Absorption and fluorescent signals will be visible from phenyl rings, tryptophan, and bound cofactors such as ATP and adenine (which is fluorescent). Positions of monoclonal antibodies on virus surfaces are now identifiable experimentally using NSOM; it should be possible to map binding sites on virus and cell surfaces using fluorescently labelled antibodies.

Single molecule detection has been proposed as a tool for rapid base-sequencing of DNA. Other optically-based cellular imaging techniques must be distinguished. Optical Coherence Tomography (OCT) uses Michelson interferometry to achieve ~10 micron spatial resolutions over tissue depths of 2-3 mm in nontransparent tissue in the near infrared.

However, OCT requires numerous physical components not easily implemented on a micron-size detector, femtosecond pulse shaping, and illumination power levels of ~10^7 watts/m^2 >> 100 watts/m^2 "safe" continuous limit in tissue. Bioluminescence techniques in which the light source is placed inside the tissue has a spatial resolution limited to ~10% of the depth, or ~10 microns resolution at a ~100 micron depth.

Coherent anti-Stokes Raman scattering (CARS) already permits organelle imaging in living cells and can in theory create a point-by-point chemical map of a cell using two intersecting lasers. Three-dimensional observation of microscopic biological nonliving structures by means of X-ray holography requires a high degree of spatial coherence and good contrast between target and surroundings.

Good contrast may be achieved in the wavelength range between the K absorption edges of carbon (l = 4.37 nm) and oxygen (l =2.33 nm), the "water window" where carbon-containing biological objects absorb radiation efficiently but water is relatively transparent.

A 50-micron diameter emitter has been tested that uses near-IR 5-femtosecond laser pulses impinging upon a helium gas target to create a well-collimated (<1 milliradian) beam of coherent soft X-rays at a 1 KHz repetition rate producing a brightness of 5×10^8 photons/mm^2-milliradian^2-sec, a peak X-ray intensity of $>10^{10}$ watts/m^2 on the propagation axis behind the He target.

Cell Volume Sensing

Many cellular parameters need not be measured directly in order to be detected by a medical nanodevice. Cell volume sensing is a case in point. An intracellular nanodevice can indirectly monitor changes in the volume of the cell in which it resides by one of two methods.

First, measurements of the mechanical deformation of the cellular membrane, stretch-activated channels, or cytoskeletal strains and structural changes are quite sensitive to alterations in total cell volume.

Second, concentration or dilution of the cytoplasmic environment through cell shrinkage or swelling leads to the activation of various volume-regulation responses which may be detected by the nanorobot.

Changes in the concentration of soluble cytosolic proteins may nonspecifically affect enzyme activity via "macro-molecular crowding". Minor changes in cell volume can cause severalfold changes in ion transport.

Cellular signalling entities that have been linked to the transduction and amplification of the primary volume signal include Ca^{++} transients, phosphoinositide turnover, eicosanoid metabolism, kinase/phosphatase systems such as JNK and p38, cAMP, and G-proteins.

These signal amplification pathways may be monitored using chemical concentration sensors aboard the medical nanodevice, allowing the nanorobot to eavesdrop on the natural sensory channel traffic of the cell.

NONINVASIVE NEUROELECTRIC MONITORING

In many ways the neuron is the most nanomedically important class of cell in the human body.

The following is a brief summary of noninvasive (i.e., no axonal membrane penetration) nanotechnological methods for monitoring the electrical traffic of individual neurons.

Noninvasive measurement of axonal traffic within nerve bundles will require multiple sensors and greater sensitivity to compensate for shielding by the perineurium, a tight resistive sheath enclosing the bundle ~20 microns thick with resistivity ~4000 ohm-cm.

ELECTRIC FIELD NEUROSENSING

The "typical" ~20 micron human neuron discharges 5-100 times per second, moving from -60 mV potential to +30 mV potential in ~10^{-3} sec. Thus the variation in electric field at the axonal surface is ± 4500 volts/m.

Since electric field sensors can detect 100 volt/m fields up to GHz frequencies, an electric sensor attached by circumaxonal cuff or pressed against the axonal surface (possibly at the node of Ranvier) should readily detect each action potential discharge. (The diameter of human nerve axons is 0.1-20 microns.)

By 1998 silicon-to-neuron extracellular junctions already permitted direct stimulation of individual nerve cells in vitro without killing the cells, and extracellular electrodes were commonly used to detect neuronal electrical activity noninvasively, both in vivo and in vitro.

Artificial electric fields may also be employed to trigger or moderate neural signals. Cell membrane capacitance is typically ~0.01 picofarads/micron^2, and varies with the state of the health of the cell.

MAGNETIC FIELD NEUROSENSING

The magnetic flux density caused by a single action potential discharge is ~0.1 microtesla at the axonal surface, which may be detected by a "compass oscillator" type magnetosensor with a ~KHz maximum sampling rate.

It might be possible for artificial magnetic fields to directly influence neural transmissions. Even a static 65 millitesla field has been shown to reduce frog skin Na^+ transport by 10-30%.

Each neuronal discharge develops an electrical energy of ~20 picojoule (~10^{10} kT), far smaller than the magnetic energy stored in a B = 1.4 tesla field of a permanent micromagnet traversing an L^3 = (20 micron)^3 volume which from Eqn. is $B^2 L^3 / 2 \mu_0$ ~ 6000 pJ. If properly manipulated, such a field may be sufficient to enhance, modulate, or extinguish a passing neural signal.

Neurothermal Sensing

While neurons come in many shapes and sizes, our "exemplar" ~14,000-micron^3 neuron discharging ~90 mV into an input impedance of ~500 Kohms produces ~0.2 microampere current per pulse and generates a continuous (average) 100-300 pW of waste heat as measured experimentally.

The discharge rate of 5-100 Hz can produce brief surges up to ~2000 pW during a high-frequency train, but the duty cycle of such trains is far less than 100%, reducing time-averaged dissipation to the observed 100-300 pW range.

Single impulses are measured experimentally to produce 2-7 microkelvin temperature spikes in cold or room-temperature mammalian non-myelinated nerve fibers and ~23 microkelvins in non-myelinated garfish olfactory nerve fibers at an energy density ranging from 270-1670 joules/m^3-impulse from 0-20°C.

In non-myelinated fibers the initial heat occurs in two temperature-dependent phases: a burst of positive heat, followed by rapid heat reabsorption. The positive heat derives from the dissipation of free energy stored in the membrane capacity, and from the decrease in entropy of the membrane dielectric with depolarization.

An L ~ 20-micron neuron in good thermal contact with an aqueous heat sink at 310 K has thermal conductance L K_t ~ 10^{-5} watts/K, so trains of 5-100 Hz impulses lasting 1 second should raise cellular temperature by 10-30 microkelvins; up to ~200 microkelvin thermal spikes from such trains have been observed experimentally.

These events are easily detectable by nanoscale thermal

sensors capable of ~1 microkelvin sensitivity up to ~1 KHz. The ~microkelvin heat signature of individual impulses or very short pulse trains can probably be temporally resolved because the minimum pulse repetition time is ~10 millisec (at 100 Hz) which is much longer than the thermal time constant for an L ~ 20-micron neuron which is $t_{EQ} \sim L^2 C_V / K_t \sim 3$ millisec.

Direct Synaptic Monitoring

The synaptic cleft between the axonal presynaptic terminal and the dendritic postsynaptic membrane is 10-20 nm in most synapses, although in the vertebrate myoneural junction it may be as large as 100 nm. Contact area per bouton is ~1 micron^2, giving a total gap volume of ~10^7-10^8 nm^3.

The density of acetylcholine receptors is highest in muscles along the crests and upper thirds of the junctional folds (~10,000/micron^2), and is lowest in the extrasynaptic regions (~5/micron^2).

Each action potential discharge triggers the release of ~10^4-10^5 molecules of acetylcholine into the gap volume of an active neuromuscular junction (diffusion time ~1 microsec), raising c_{ligand} from near zero to ~3×10^{-4} molecules/nm^3 (~0.0005 M) in ~1 millisec, followed by near-complete hydrolyzation by acetylcholinesterase during the 1-2 millisec refractory period.

Into the gap volume may easily be inserted a ~10^5 nm^3 neurotransmitter concentration sensor able to measure ~100 acetylcholine molecules in ~1 millisec, thus detecting pulses at the fastest discharge rate.

A similar device could be used to precisely regulate neurotransmitter concentration at the junction, and hence the neural signal itself, under nanodevice control. Simple electrochemical and mechanochemical artificial synapses have been demonstrated.

CELLULAR RF AND MICROWAVE OSCILLATIONS

Starting in 1968, H. Frohlich, observing that millivolt electrical potentials maintained across cell membranes ~10 nm thick give rise to huge fields ~10^7 volts/m possibly producing an electret state, theorized that membrane molecules must be highly electrically polarized and thus could interact to produce coherent surface acoustic vibrational modes in the 10-100 GHz (microwave) frequency range; the longest wavelength is about twice the membrane thickness.

Interestingly, this frequency span is very close to the maximum trigger/reset frequency for bioelectronic molecules. Note that (~100 mV) (1.6 x 10^{-19} coul) ~ 4 kT, so a membrane molecule with a single charge on either end should be reliably reoriented by a depolarization wave, coupling pressure waves and electrostatic field fluctuations.

However, the direct detection of 10-100 GHz millimeter radiation by non-nanotechnological means is experimentally difficult and controversial because the tests must be performed in vivo in close proximity to an actively metabolizing cell in water — and water strongly absorbs microwaves over macroscale ranges.

Nevertheless, active cells have shown enhanced Raman anti-Stokes scattering, an effect ascribed to the converse of the Frohlich oscillations.

In one study, the normalized growth rate of yeast cultures was enhanced or inhibited when irradiated by CW microwave fields of ~30 watts/m^2 of various frequencies; growth rate data spanning 62 separate runs revealed a repeatable frequency-dependent spectral fine structure with six distinct peaks of width ~10 MHz near 42 GHz.

Investigations of related phenomena are ongoing and voluminous; the interested reader should peruse *Bioelectromagnetics*, the archival journal of this field. 100 GHz waves attenuate only ~1% after passing through ~3 microns of soft tissue.

A single electron injected into an integral membrane protein could act as an oscillating dipole, making a 300 volt/m signal 10 nm from the protein antenna with an energy transfer of ~0.004 kT per cycle; ~1000 oscillating electrons could produce a measurable field.

A 20-micron diameter cell modeled as a nonuniform spherical dipole layer with transmembrane dipoles located 10 nm apart and embedded in a dissipative medium could produce $10^2 - 10^5$ volt/m microwave fields 1-10 microns from the cell surface.

Thus, a variety of rf and microwave electromagnetic emanations may in theory be detectable both within and nearby living cells which could prove diagnostic of numerous internal states.

Such states may include cytoskeletal dynamics, metabolic rates, plasmon-type excitations due to the collective motion of ions freed in chemical reactions, positional, rotational or conformational changes in biological macromolecules and membranes, internal movements of organelles and nerve traffic conduction, cellular pinocytosis, cellular reproduction events, cell membrane identity, and cell-cell interactions.

MACROSENSING

Macrosensing is the detection of global somatic states (inside the human body) and extrasomatic states (sensory data originating outside the human body). While the treatment here is necessarily incomplete, the discussion nevertheless gives a good feel for the kinds of environmental variables that internally-situated nanodevices could sense.

Not all capabilities outlined here need be available on every nanorobot, since injection of a cocktail of numerous distinct but mutually cooperative machine species allows designers to take full advantage of the benefits of functional specialization. In many cases, a given environmental variable can be measured by several different classes of sensor device. However, since these devices are microscopic it is in theory possible to operationalize almost all of the macrosensing capabilities described below in one patient using just a billion devices (~10 mm^{-3} whole-body deployment density), a total volume of ~1 mm^3 of nanorobots or ~0.1% of the typical ~1 cm^3 therapeutic dose.

CYTO-AUSCULTATION

Can sounds generated by a single cell be detected, and thus be useful for diagnosis? Probably not, given that low frequency acoustic radiators are notoriously inefficient.

For example, mitochondrion organelles of the giant amoeba *Reticulomyxa* are shuttled back and forth by 1–4 cytoplasmic dynein motors while riding on the outside of a bundle of 1–6 microtubules.

Each dynein motor generates 2.6 pN of force and drives the mitochondria at up to ~10 micron/sec, developing 0.3–1.0 × 10^{-16} watts of mechanical power within each 320-nm diameter organelle.

Taking each organelle as a cylindrical acoustic radiator with a mechanical input power of P_{in} ~ 10^{-16} watts at ν ~ 1000 Hz, the output acoustic pressure at the organelle surface is only ~10^{-9} atm, an acoustic power intensity I ~ 10^{-15} watts/m^2 which is not detectable by micron-sized nanorobots.

Nevertheless, cells and intracellular elements are capable of vibrating in a dynamic manner with complex harmonics that can be altered by growth factors and by the process of carcino-genesis, so the possibility cannot be completely ruled out.

Blood Pressure and Pulse Detection

Blood pressure ranges from 0.1– 0.2 atm in the arteries to as low as 0.005 atm in the veins. The systolic/diastolic differential ranges from 0.05–0.07 atm in the aorta and 0.01–0.02 atm in the pulmonary artery, falling to 0.001–0.003 atm in the microvessels, or 0.003–0.005 atm if the precapillary sphincter is dilated.

In venous vessels, pulse fluctuations are 0.002–0.010 atm in the superior vena cava, 0.004 – 0.006 atm in the subclavian vein, ~0.004 atm in venules generally, and ~0.0005 atm in the brachial vein.

There is also a ~0.05 Hz random fluctuation in the microvessels with amplitude on the order of 0.004 – 0.007 atm. Both blood pressure and pulse rate can be reliably monitored by a medical nanodevice virtually anywhere in the vascular system using a (68 nm)^3 pressure sensor with ~0.001 atm sensitivity.

Pulse propagation through body tissue is somewhat muted due to absorption in compressible fatty membranes, but most cells lie within 1 – 3 cell-widths of a capillary so the cardiac acoustic signal should still be measurable using more sensitive detectors.

The time-averaged interstitial pressure in subcutaneous tissue is 0.001–0.004 atm. Arterial pulse waves (vascular oscillations) carry subtle messages about the health of internal organs and the arterial tree.

The idea of using pulse waves for diagnosis has a long history dating back 2000 years in China.

Waves detected by manual probing are classified using such subjective and qualitative descriptors as floating, deep, hidden, rapid, slow, moderate, feeble, replete, full, thready, faint, weak, soft, slippery, hesitant, hollow, firm, long, short, swift, running, intermittent, uneven, τ t, string-tight, gigantic, or tremulous.

Abnormal waves were empirically related to disease states. Wave data gathered by nanodevices could make possible a theoretically sound, quantitative system of noninvasive observation, classification, and diagnosis as a supplement to other nanomedical tools.

Respiratory Audition

The variation of mechanical pressure over a complete respiratory cycle is ~0.003 atm in the pleura, ~0.002 atm at the alveoli, detectable by nanomedical pressure sensors positioned in the vicinity of the respiratory organs.

Holding a deep breath further stretches the pulmonary elastic tissue, up to 0.02 atm. However, turbulent flows at Reynolds number $N_R > 2300$ in the trachea, main bronchus and lobar bronchus produce a whooshing noise that may be the loudest noncardiac sound in the human torso during conventional auscultation.

The energy dissipation for turbulent flow in a tube is:

$$P_{turb} = P_{lam} Z = 8 \pi_{air} v^2 L Z$$

(watts)

where P_{lam} is the dissipation for laminar (Poiseuillean) flow in a long circular cylindrical tube of length L, v is the mean flow velocity, and turbulence factor $Z = 0.005 (N_R^{3/4} - (2300)^{3/4})$, a well-known empirical formula.

For $\eta_{air} = 1.83 \times 10^{-5}$ kg/m-sec for room-temperature air (20°C), $P_{turb} = 0.87$ milliwatts for the trachea; $P_{turb} = 0.66$ milliwatts for the main bronchus (L = 0.167 m, v = 4.27 m/sec and $N_R = 3210$); and $P_{turb} = 0.09$ milliwatts for the lobar bronchus (L = 0.186 m, v = 4.62 m/sec and $N_R = 2390$), totalling ~1.6 milliwatts acoustic emission from a ~120 cm³ upper tracheobroncheal volume.

This is a power density of 13 watts/m³ corresponding to a pressure of 4×10^{-5} atm assuming a 300 millisecond measurement window at the maximum respiration rate. The amplitude of an acoustic plane wave propagating through tissue attenuates exponentially with distance due to absorption, scattering and reflection.

The amplitude is given approximately by:

$$A_x = A_D e^{-\alpha F x}$$

where A_0 is the initial wave amplitude in atm, A_x is the amplitude a distance x from the source, and α is the amplitude absorption coefficient.

The function F expresses the frequency dependence of the attenuation. For pure liquids, $F = F_{liq} = \nu^2$ (Hz²); for example, $\alpha_{liq} = 2.5 \times 10^{-14}$ sec²/m for water at room temperature. However, for soft tissues, $F = F_{tiss} \sim \nu$ (Hz).

Mechanical Body Noises

Many other mechanical body noises should be globally audible to properly instrumented medical nanodevices. If normal chewing motions (of hard foods) release 1–10 milliwatts in a ~100 cm³ oral volume with a ~1 sec jawstroke, power density is ~10 watts/m³ or ~10^{-4} atm of toothcrunching noise. A stomach growl registering 45 dB (vs. 30 dB whisper, 60 dB normal conversation) at 2 meters has a source power of 160 milliwatts; released from a 10 cm³ gastric sphincter volume gives a ~2×10^{-6} atm acoustic wave, detectable throughout the body.

Walking and running releases 20-100 joules/footfall for a 70 kg man; assuming the energy is absorbed within a ~1 cm thickness or within ~1 second by the sole of the foot, Equation. implies an upward-moving planar compression wave of 0.4-2.0 atm, easily detectable by acoustically instrumented nanodevices body-wide. Hand-clapping generates 0.02-0.2 atm pulses, also easily detectable.

Lesser noises including ~30 millisec hiccups at (4–60)/min, intestinal and ureteral peristalsis, sloshing of liquid stomach contents, heart murmurs, a tap on the shoulder by a friend, nasal sniffling and swallowing, clicks from picking or drumming fingernails, crepitations, manustuprations and ejaculations, the rustling noise of clothing against the skin, flapping eyelids, anal towelling, bruits (including murmurs and thrills) due to vascular lesions, dermal impact of water while showering, copulatory noises, urethral flow turbulence during urination, transmitted vibrations from musical instruments, creaking joints, and squeaking muscles can be detected locally if not globally.

Implantation of significant interconnected in vivo diamondoid structures may produce increased sensitivity to internal noises, due to the extremely low acoustic absorption coefficient of diamond.

Vocalizations

Average source power for conversational speech in air is ~10 microwatts at the vocal cords (60 dB), up to ~1000 microwatts for shouting (90 dB) and as little as 0.1 microwatts (30 dB) for whispering. Vocal cord surface area ~1 cm^2, giving an acoustic intensity I ~ 0.001–10 $watts/m^2$. (Using the decibel notation, dB = $10 \log_{10} (I/I_0)$, where $I_0 \sim 5 \times 10^{-13}$ $watts/m^2$ in air, $I_0 \sim 1 \times 10^{-16}$ $watts/m^2$ in water.) In a planar traveling wave, pressure amplitude A_p (N/m^2) is related to power intensity I by:

$$A_p = (2\rho \nu_{sound} I)^{1/2}$$

(N/m^2)

For water at 310 K, ρ = 993.4 kg/m^3 and v_{sound} = 1500 m/sec, therefore A_p = 0.0005–0.05 atm for speech, detectable by nanodevices throughout the body due to minimal attenuation at audible frequencies.

Other easily detectable vocalizations include whistling, humming, coughing, sneezing, rales, wheezing, expectorating, eructations, flatus, vomiting, hawking and noseblowing.

ENVIRONMENTAL SOURCES

Can in vivo nanodevices directly detect sounds emanating from the environment outside of the body, such as other people talking in the same room or a door slamming? The waves from an external acoustic source of power P_R watts travel through the air and, upon arriving at the air/skin interface a distance x_R from the source with amplitude $A_{incident}$, are transmitted through the interface with amplitude $A_{transmit}$.

For a specular reflector — interface dimensions (human body ~ 2 m) > acoustic wavelength (~0.03–3.0 m for typical audible sounds in air) — with acoustic impedance Z_1 and Z_2 on either side of the interface and perpendicular incidence,

$$A_{transmit} = A_{incident}\left[1 - \mathrm{abs}\left(\frac{Z_1 - Z_2}{Z_1 + Z_2}\right)\right]$$

Acoustic impedance, like the speed of sound, is essentially frequency-independent over the nanomedically-relevant range of ultrasonic frequencies. For Z_{air} = 400 kg/m^2-sec and assuming Z_{skin} ~ 1.6 × 10^6 kg/m^2-sec, then $A_{transmit}$ = (5 × 10^{-4}) $A_{incident}$.

In other words, there is ~99.95% reflection from the air-skin interface, which is why coupling mediums like gels and oils are commonly employed in ultrasound imaging. If an immediately subdermal nanodevice can detect a minimum $A_{transmit}$ ~ 10^{-6} atm, then from simple geometry:

$$P_R = \frac{2\pi \times_R^2 A_{incident}^2}{\rho v_{sound}}$$

For STP (1 atm, 0°C) air, ρ = 1.29 kg/m^3 and v_{sound} = 331 m/sec. If the minimum detectable pressure ~10^{-6} atm ~$A_{transmit}$, then at a distance of x_R = 2 meters the acoustic source must have a power of ~2000 watts, far exceeding the ~1 milliwatt output of a person loudly shouting.

To hear normal conversation at x_R = 2 m, minimum

nanodevice detector sensitivity falls to 7×10^{-11} atm requiring a subdermal pressure nanosensor ~(17 micron)^3 in size, roughly the dimensions of a single human cell; other methods may prove more efficient.

Of course, an ex vivo acoustic nanosensor may receive sound that has passed through no interface, hence may detect pressure waves ~3 orders of magnitude lower in amplitude. Assuming $A_{transmit} = A_{incident}$, $P_R \sim 600$ microwatts, so ex vivo nanorobots with a 0.3 micron^3 sensor could hear people shouting at $x_R = 2$ meters.

To hear talking (~10 microwatt source) requires a 2.4 micron^3 ex vivo pressure sensor (limit ~10^{-7} atm). Optimally positioned and calibrated nanomedical pressure sensors could directly measure changes in the ambient barometric pressure to within $\pm 10^{-6}$ atm.

Normal atmospheric variation due to weather ranges from 0.94–1.05 atm; such slow moving changes are readily monitored.

Very near the Earth's surface, the air pressure P at altitude h above sea level is approximated by $P = e^{-k_p h}$ (atm), where $k_p = 1.16 \times 10^{-4}\ m^{-1}$ at 20°C; at sea level, a 10^{-6} atm change in pressure reflects a change in altitude of only ~1 cm.

However, the opening or closing of a door inside a (~5 m)^3 room that displaces >125 cm^3 of air also causes a minimally detectable >10^{-6} atm pressure pulse.

Other sources of environmental pressure variation such as infrasonic (~0.2 Hz) microbaroms from offshore ocean storms, wind entering through open windows, forced-air currents from central heating or A/C systems, or even the movements of nearby people and pets may be detectable and thus may further confuse the measurement, reducing absolute accuracy unless suitable corrections are made.

KINESTHETIC MACROSENSING

Using a navigational transponder network, a population of ~10^{11} nanodevices each spaced an average ~100 microns apart throughout the body tissues can determine relative location to a positional accuracy of ~3 microns and an angular accuracy of ~2 milliradian, with the data updatable once every millisecond.

Thus dispersed, the network can continuously monitor and record the relative positions of all limbs with worst-case (cumulative error) ~0.8 mm accuracy over a 2-meter span. These devices can prepare high-resolution dynamic maps of body position, velocity, acceleration and rotation, allowing precise real-time digital kinesthesia during sports or artistic activities such as gymnastics, pole vaulting, or ballet; during transportative activities such as driving cars around hairpin turns, roller-coaster rides, military aircraft maneuvers and space launches; during precision tool-using such as needle-threading, antique watch repair, or while using the fingers as measurement calipers; during self-defence activities requiring complex motions such as karate or judo; and during emergency situations such as automobile crashes and tumbling motions during falls from great heights.

All such sensory data is readily outmessaged to the patient or user in real time. Nanorobots could assist in path integration or the reconstruction of experienced limb trajectories. Given the ability to determine rotation rates and to measure applied forces and torques, the network should also be able to forecast the anticipated future positions of limbs.

By comparing these projections to actual results, the network can then infer the viscosity of the medium in which the activity is taking place, whether the environment is stable or is translating or rotating in some direction, and whether the human user is physically supported or in free-fall.

Internal nanodevices can directly measure if a body is sitting, standing, laying prone, inverted, falling, floating or diving, and then communicate that information directly to the human user.

Detection of patient activity states — e.g., the patient is sitting, standing, or walking — can be used to control nanorobot behaviours, activate or deactivate outmessaging displays, and so forth. A simple macroscale wearable tactile "compass belt" that would convey desired geographical directions has already been proposed.

Orientational Macrosensing

Since nanopendular sensors can reliably determine the direction of the local gravity field vector to within ~2 milliradians in ~10^{-4} sec, the navigational network can poll its members and arrive at an accurate consensus on which direction is up.

Nanodevices affixed to relatively stable hard body parts will exhibit more consistent orientational readings. Measurement of the gravity vector allows the vertical orientation of the rest of the body to be precisely fixed in space, especially useful for gymnasts, trapeze artists, underwater divers in murky lakes, and vestibular-impaired individuals, to whom this information may be outmessaged directly.

The gravitational vector may also be indirectly measured, albeit more slowly and less accurately, by monitoring the body's natural reactions to changes in the axis of gravitational loading. For instance, the variation in head-to-toe hydrostatic pressure for a 1.7 m tall, 70 kg human standing in a 1-g gravity field is ~0.17 atm; the ability to measure a systematic cross-body differential of 10^{-6} atm allows the detection of a ~10^{-5} g change along the lengthwise aspect, or ~10^{-4} g change along the transverse aspect.

Thus a continuously updated whole-body barostatic map allows the human body to serve as a three-dimensional gravity/orientation sensor.

Body Weight Measurement

A network of nanodevices can inventory the volume and density of each of the body's ~10^8 $(1\ mm)^3$ voxels using a combination of acoustic ranging, somatic mapping, and flowmetry. Each voxel is identified as fat, muscle tissue, bone mass, interstitial fluid, and so forth. These measurements allow the body's weight to be precisely computed as volume times density, rather than the usual method of determining weight using gravitational force data.

The existence of a natural physiological "ponderostat," a crudely analogous humoral body-mass detector, has been proposed; indeed, the insulin-leptin system has been found to serve a related purpose, and is easily eavesdropped by medical nanorobots.

Gravitational Geographic Macrosensing

Medical nanodevices can measure variations in the gravity field to ~10^{-6} g's for L = 20 micron gravimeters in a measurement time t_{meas} = 2–9 millisec. This implies that in vivo nanodevices can take precise measurements of their latitude and altitude relative to sea level ~100 times every second.

Gravity increases toward the poles and at lower altitudes. Specifically, using the formula of Cassinis (accounting for rotational and polar flattening effects on the Earth) with the Bouguer correction to the free air variation by altitude (assuming flat topography), measured gravity g_{meas} is given approximately by,

$$g_{meas} g_0 [1 + k_1 \sin^2(\theta_L) - k_2 \sin^2(2\theta_L)] - k_3 h + k_4 h \rho_{earth}$$

where θ_L = terrestrial latitude (equator = 0°), h = height above sea level in meters, g_0 = 9.78039 m/sec^2 (equatorial sea-level value ofg), $k_1 = 5.2884 \times 10^{-3}$, $k_2 = 5.9 \times 10^{-6}$, $k_3 = 3.086 - 10^{-6}$ sec^{-2}, $k_4 = 4.185 - 10^{-7}$, and ρ_{earth} = 5522 kg/m^3.

Since sea-level g varies from 9.78039 m/sec^2 at the equator to 9.83217 m/sec^2 at the north pole, a 20-micron gravimeter ($\Delta g = 10^{-6}$ g) detects a change in position of 1 arcmin of latitude or ~1900 meters north/south along the Earth's surface. Similarly, since at 45° latitude g varies from 9.806 m/sec^2 at sea level to 9.803 m/sec^2 at 1000 meters altitude, a 20-micron gravimeter detects a change in altitude of ~3.3 meters.

For comparison, in 1998 high-quality commercial gravity gradiometers measured gradients of ~10^{-9} g/meter and allowed the compilation of micro-g (~1 milligal) resolution aerial gravity maps; atom interferometers measured the gravitational acceleration of atoms to a precision of 10^{-10}.

To achieve such phenomenal positional accuracies, the nanodevice must be able to computationally resolve several complicating factors. First, localized mass concentrations representing nonuniformities in crustal density produce residuals of up to ±0.0006 m/sec^2, which may be removed from the data using a standard map of known terrestrial isostatic variations and anomalies.

Indeed, matching observations to such a map could provide useful longitudinal information as well. Another complication is the variation in gravity due to tidal forces amounting to ~3×10^{-7} g's, twice daily, which lies at the limits of detectability for a 20-micron gravimeter.

Other minor geodesic and terrain-related corrections, too complicated for discussion here, may also need to be applied in certain circumstances. Note that the presence of nearby heavy objects does not influence measurement accuracy: a 100-ton building 10 meters away adds a lateral acceleration of only 7×10^{-9} g's to a human body.

One final complication is that patient movements create kinematic accelerations that must be distinguished from the gravitational accelerations.

Gravity readings can be corrected by taking derivatives of the signals from kinesthetic monitoring to give gravity in the reference frame of the patient's room, and many individual measurements may be averaged to improve accuracy since the gravity vector normally changes only very slowly.

VASCULAR-INTERSTITIAL CLOSED ELECTRIC CIRCUITS

In vivo studies of the electric properties of blood vessels shows that the walls of veins and arteries present a specific electric resistance ~200 times greater than the conductive medium of blood (plasma) — roughly 200 ohm-m vs. 0.7 ohm-m. Thus blood vessels are properly regarded as relatively insulated conducting cables which can electrically connect an injured tissue with surrounding noninjured tissue. Capillaries form an electric junction with the interstitial fluid, so the electric gradient can be canceled by ionic transport.

Cell membranes are insulating dielectrics with resistance and capacitance, penetrated by ionic channels or gates for ionic transport. Additionally, in 1941 Szent-Gyorgyi suggested the possibility of semi-conduction in proteins, a theory which has since been elaborated by many investigators.

B. Nordenstrom has proposed that the above schema represents a system for selective electrogenous mass transport of material between blood and tissue which he calls Vascular-Interstitial Closed (electric) Circuits (VICCs) — in effect, an additional electro-circulatory system operating in parallel with the well-known diffusive, osmotic, and hydrodynamic mechanisms of regular bloodstream materials transport.

As long as ions can leak through open pores and migrate through ion channels, long distance transport cannot take place and the VICC system remains primarily a local (e.g., within tissues), not global, electrical circuit.

Connections also exist with conductive media including cerebrospinal fluid, bile, and urine. Nanorobots capable of monitoring the status of local VICCs may rapidly and efficiently acquire a wealth of systemic information without the need for direct inspection of the affected tissues.

For example, when a working muscle produces catabolic products such as lactic acid, an electrochemical potential gradient develops between the muscle and surrounding tissue in a known manner. Injured tissues become polarized in relation to surrounding tissue, as initial catabolic degradation products acidify the tissue.

Malignant neoplasms, benign neoplasms, and internally necrotic granulomas all may polarize electrically in relation to surrounding tissue; a vascular thrombus also contains ionized material. Large deflections in the diffusion potential of blood occur if blood is deoxygenated or is infected with Gram-negative bacteria.

Blood changes its electric potential from +500 mV to +1000 mV during spontaneous coagulation, and the spontaneous electric potential at the site of a crush injury of the iliac crest in rats oscillated several times between +190 mV and –60 mV over a four day experimental trial.

Nordenstrom proposes that in vivo electrophoresis via the VICC system may also play a role in mediating leukotaxis. When tissue is artificially polarized by electrodes, accumulation of granulocytes with margination and development of pseudopods is the result.

In one experiment, exposure of a mesenterial membrane to a 1 microampere 1 volt field for 30 minutes stimulated diapedetic bleeding near the anode. At higher power, the arterio-capillaries contracted, narrowed, and emptied of blood cells, while the veno-capillaries and venules widened and filled with granulocytes.

Monitoring such actions of local VICCs experiencing natural voltage fluctuations could alert medical nanorobots to the presence of local injuries, tissue changes, or pathologies that otherwise might go unnoticed for a considerable time.

Electric/Magnetic Geographic Macrosensing

Detection of electric fields to ~100 volts/m, the normal vertical atmospheric gradient, in theory might allow determination of aboveground altitude to ~1 m accuracy if the time between recalibrations is very brief. However, a human body which is a good conductor, standing on level ground, acquires a slight negative surface charge and becomes "part of the ground" electrically.

This distorts upward the equipotentials that usually run parallel to the ground, thoroughly corrupting an altitude measurement. Nearby lightning hits and storms are detectable, which is useful.

However, medical nanodevices with appropriate magnetometers can measure variations in magnetic field to ~0.1 microtesla using a $(660\ nm)^3$ permanent magnet sensor, in a measurement time t_{meas} ~ 0.6 millisec.

Geomagnetic field maps of the Earth's surface are highly nonisotropic in both latitudinal and longitudinal directions; knowledge of the field plus the absolute direction of true north gives longitude information from the separation of the planetary magnetic and spin poles. This implies that properly equipped in vivo nanodevices can establish their latitude and longitude on Earth's surface ~1000 times every second by this means. In particular, the horizontal component of the geomagnetic field vector ranges from 0–41 microtesla from magnetic pole to magnetic equator; the independent vertical field component ranges from 0–70 microtesla.

Thus a 0.1-microtesla sensor should resolve each component to an accuracy of ~11 arcmin or ~20 km in both latitude or longitude. Terrestrial surface magnetic anomalies (e.g., iron deposits) produce local variations of 0.03–30 microtesla and are thus detectable.

Additional complications to these measurements include artificial magnetic field sources, the 0.01–0.1 microtesla solar daily variation, occasional magnetic storms producing erratic geomagnetic fluctuations of 0.01–5 microtesla, sudden-commencement ionospheric electrojets up to 0.3 microtesla, and various long-term secular variations in the geomagnetic field. Radio, television, and direct broadcast satellite signals probably are not detectable by individual nanodevices, but such detection may be indirectly achieved in vivo using nanorobot-accessible dedicated macroscale antennas implanted in the body.

Piezoelectric Stress Macrosensing

The piezoelectric effect is "the production of electrical polarization in a material by the application of mechanical stress". Many materials in the human body are piezoelectric, including tendon and elastin, dentin and bone. This polarization, or surface charge, varies as the physical stress imposed on the material changes over time.

Thus, measurement of the piezoelectric effect in bone or tendon can provide information on the physical loads that are being carried by these materials. For example, a shear stress applied along the long axis of a bone alters the polarization voltage that appears on a surface at right angles to the axis. In another experiment, piezoelectric surface charges on a loaded human femur were measured to range from –131 to +207 picocoulombs/cm^2 (–8 to +13 charges/micron^2) from one end of the bone to the other end, depending upon position along the shaft.

Real-time monitoring of this data would permit specific inferences as to the amount of load the bone was carrying and from what direction, including bending, shearing, and twisting forces, from which whole-body activity states could subsequently be inferred. Knowledge of these surface charge variations might also be exploited in osteographic or functional navigation.

Optical Macrosensing

In vivo medical nanodevices can gather little useful optical information from the external environment. Here's why. In biological soft tissues, scattering dominates absorption except in the pigmented layers of the epidermis and stratum corneum.

Thus the propagation of light in tissues may be regarded as occurring in two steps. In the first step, optical photons of intensity I_0 falling perpendicularly on skin are transmitted according to Beer's law through tissue to a depth z with a transmitted intensity of,

$$I_z = I_0(1 - r_{sp})e^{\sigma_t z}$$

(watts/m^2)

where r_{sp} = specular reflection coefficient for visible light (Fresnel reflection at the air-tissue surface) ~ 4%–7%, or 0% if the original light source lies within the body; and the transmission coefficient $\sigma_t = \sigma_a$ (absorption coefficient ~ 300 m^{-1}) + σ_s (scattering coefficient ~30,000 m^{-1}) for various human soft tissues at optical wavelengths. (Coefficient values for the most heavily pigmented skin layers may be 5–7 times higher.)

Typical values for σ_t ~ 10,000–100,000 m^{-1}, average ~30,000 m^{-1} for soft tissue, although exceptionally clear tissues with σ_t = 1000 m^{-1} have been reported. Thus the mean free path of an optical photon in human soft tissue is 10–100 microns, average ~ 30 microns (~1.5 tissue cell-widths), up to an extreme maximum of ~1 mm for the most transparent tissues known. For σ_t ~ 30,000 m^{-1}, at z = 150 microns, I_z/I_0 ~ 0.01 and ~99% of all photons have been scattered at least once from their initial path. In the second step, in an unbounded medium the patch of fully scattered photons continues to propagate through the tissue via diffusion until all photons are absorbed. The complicated governing diffusion equation has not yet been been completely solved analytically, but the asymptotic diffusive fluence is given roughly by,

$$I_d - I_1 e^{\sigma_d z}$$

(watts/m^2)

where σ_d is the diffusion exponent or effective attenuation coefficient (averaging ~900 m^{-1} for typical soft tissues but with an extremely wide range reported over the ultraviolet, visible, and near-infrared wavelengths, from 10–1,000,000 m^{-1}.) As a crude approximation, the initial intensity of the fully scattered photon patch $I_i \sim I_0 (1 - r_{sp}) a_{scat}$, where $a_{scat} = \sigma_s / (\sigma_s + \sigma_a) \sim 0.987$ is the albedo for single particle scattering.

Over the optical band from 400–700 nm, incident intensity I_0 = 100–400 watts/m^2 when standing in direct sunlight; artificial lighting in homes and offices is typically 0.1–10 watts/m^2; moonlight provides only $I_0 \sim 10^{-4}$ watts/m^2; and the absolute threshold for human vision is ~10^{-8} watts/m^2.

For $\sigma_d \sim 900$ m^{-1}, the intensity ratio of transmitted/incident visible light falls to $I_d / I_0 = 0.1$ at z = 2.5 mm depth (~eyelids), ~10^{-4} at z = 1 cm depth, and ~10^{-11} at z = 2.8 cm — a depth at which the tissue would be completely dark to the human eye even under direct sunlight illumination at the outermost skin surface. An optical nanosensor with N_{sensor} receiver elements, with each receiver element having an area A_e = 1 nm^2 and capable of single-photon detection, requires an integration time for reliable detection of e^{SNR} photons given by,

$$t_{meas} - \frac{hve^{SNR}}{I_d N_{sensor} A_e}$$

where h = 6.63 × 10^{-34} joule-sec and n = 4.3–7.5 × 10^{14} Hz for optical photons. Let us assume that a patient is standing in $I_0 \sim 400$ watt/m^2 direct unfiltered sunlight, and that t_{meas} = 1 sec, SNR = 2, N_{sensor} = 25,000 elements giving a nanorobot eyespot of $A_e N_{sensor}$ = 0.025 micron^2, and that σ_d = 900 m^{-1}. Then the maximum tissue depth for optical photon detection is $z_{max} \sim$ 17 mm ($I_d \sim 10^{-4}$ watt/m^2) which includes a volume of tissue comprising up to ~30% of total body volume.

Changes in illumination at the minimum indoor artificial level of ~0.1 watts/m^2 are visible to $z_{max} \sim 7$ mm depth. This eyespot, if exposed to air on the outermost surface of the skin, is just sensitive enough to detect full moonlight.

Our general conclusion is that variations in normal indoor lighting may be directly measurable by nanorobots stationed within the outermost ~1 cm of body tissues. Imaging, as opposed to the simple illumination detection system described here, is a much more difficult design challenge.

NEURAL MACROSENSING

The ability to detect individual neural cell electrical discharges noninvasively in many different ways, coupled with the abilities (A) to recognize and identify specific desired target nerve cells and (B) to pool data gathered independently by spatially separated nanodevices in real time, offers the possibility of indirect neural macrosensing of complex environmental stimuli by eavesdropping on the body's own regular sensory signal traffic.

Nanomonitors positioned at the afferent nerve endings emanating from hair cells located in the otolithic membrane of the utricle and saccule and in the cristae ampullaris of the semicircular canals allow medical nanodevices to directly record, amplify, attenuate, or modulate the body's own sensations of gravity, rotation, and acceleration, although kinesthetic sensory management might also be required for complete control.

Motor neurons likewise can be monitored to keep track of limb motions and positions, or specific muscle activities, and even to exert control. Neuron-resident nanorobots may detect auditory effects induced in the brain by pulsed microwaves from external sources. Feline cochlear neurons sensitive to audible frequencies of >300 Hz respond to single microwave pulses at a threshold-specific absorption rate of 6–11 watts/kg-pulse, while in humans the threshold for effect, which depends on energy per pulse, may be as low as ~0.02 joule/m^2-pulse for people with low hearing threshold.

Olfactory and gustatory sensory neural traffic similarly may be eavesdropped by nanosensory instruments. Nerve taps in the medulla oblongata or at the phrenic nerve that drives the diaphragmatic muscles allow direct monitoring of respiratory activity.

Pain signals may be recorded or modified as required, as can mechanical and temperature nerve impulses from other receptors located in the skin. Even psychological variables such as emotionality, vigilance, and mental workload may be directly monitored by measuring ANS activity in sympathetic efferent fibers outside of the brain.

The most complex and difficult challenge in neural macrosensing will be optic nerve taps. The retina is thoroughly vascularized, permitting ready access to both photoreceptor and integrator neurons.

However, the optic nerve bundle itself has ~10^6 tightly bunched individual nerve fibers, a 10–100 MHz signal bandwidth, and significant natural data compression techniques which all must be untangled in real time. Developing algorithms capable of interpreting raw optical nerve traffic, say, to recognize a specific human face or a specific scene in the vision field, would prove a significant research challenge.

Direct monitoring of photo-receptors or retinal membrane potentials and other techniques may simplify untangling of the signal compression. Rapid visual field identification using artificial neural nets is also a subject of much current research interest. Eyeball rotations, eyelid position, pupil aperture, and lens accommodation under ciliary muscle control must be monitored to supplement the vision field analysis.

Chapter 5

Features

USES

Disease and ill health are caused largely by damage at the molecular and cellular level. Today's surgical tools are, at this scale, large and crude. From the viewpoint of a cell, even a fine scalpel is a blunt instrument more suited to tear and injure than heal and cure.

Modern surgery works only because cells have a remarkable ability to regroup, bury their dead and heal over the injury. Nanotechnology, "the manufacturing technology of the 21st century," should let us economically build a broad range of complex molecular machines (including, not incidentally, molecular computers).

It will let us build fleets of computer controlled molecular tools much smaller than a human cell and built with the accuracy and precision of drug molecules. Such tools will let medicine, for the first time, intervene in a sophisticated and controlled way at the cellular and molecular level.

They could remove obstructions in the circulatory system, kill cancer cells, or take over the function of subcellular organelles. Just as today we have the artifical heart, so in the future we could have the artificial mitochondrion.

Equally dramatic, nanotechnology will give us new instruments to examine tissue in unprecedented detail. Sensors smaller than a cell would give us an inside and exquisitely precise look at ongoing function.

Tissue that was either chemically fixed or flash frozen could be analysed literally down to the molecular level, giving a completely detailed "snapshot" of cellular, subcellular and molecular activities.

There is broad agreement (though not consensus) that we will at some point in the future be able to inexpensively fabricate essentially any structure that is consistent with chemical and physical law and specified in molecular.

The most direct route to achieving this capability involves positioning and assembling individual atoms and molecules in a fashion conceptually similar to snapping together LEGO blocks. By designing and building programmable self replicating manufacturing systems that incorporate these principles we should be able to achieve very low manufacturing costs.

While the design and development of such programmable self replicating molecular manufacturing systems will be a major task and will likely require many years or a few decades, it appears that this kind of capability, to quote Feynman.

Design concepts for general purpose self replicating manufacturing systems have been discussed for many years, and their utility in manufacturing has been emphasized recently. These proposals draw on a body of work started by von Neumann. A wide range of methods have been considered. The von Neumann architecture for a self replicating system is the ancestral and archetypal proposal.

ARCHITECTURE FOR MANUFACTURING SYSTEM

Von Neumann's proposal consisted of two central elements: a universal computer and a universal constructor . The universal computer contains a programme that directs the behaviour of the universal constructor.

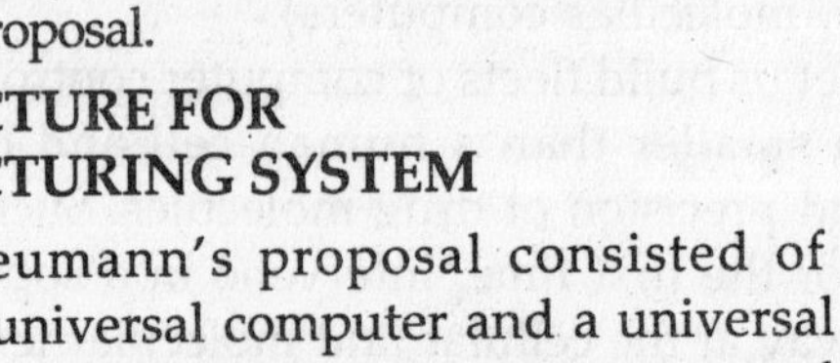

The universal constructor, in turn, is used to manufacture both another universal computer and another universal constructor. Once construction is finished the programme contained in the original universal computer is copied to the new universal computer and programme execution is started.

Von Neumann worked out the details for a constructor that worked in a theoretical two-dimensional cellular automata world (parts of his proposal have since been modeled computationally). The constructor had an arm which it could move about and which could be used to change the state of the cell at the tip of the arm.

By progressively sweeping the arm back and forth and changing the state of the cell at the tip, it was possible to create "objects" consisting of regions of the two-dimensional cellular automata world which were fully specified by the programme that controlled the constructor.

While this solution demonstrates the theoretical validity of the idea, von Neumann's kinematic constructor (which was not worked out in such detail) has had perhaps a greater influence, for it is a model of general manufacturing which can more easily be adapted to the three-dimensional world in which we live. The kinematic constructor was a robotic arm which moved in three-space and which grasped parts from a sea of parts around it. These parts were then assembled into another kinematic constructor and its associated control computer.

An important point to notice is that self replication, while important, is not by itself an objective. A device able to make copies of itself but unable to make anything else would not be very valuable. Von Neumann's proposals centered around the combination of a universal constructor, which could make anything it was directed to make, and a universal computer, which could compute anything it was directed to compute.

It is this ability to make any one of a broad range of structures under flexible programmatic control that is of value. The ability of the device to make copies of itself is simply a means to achieve low cost, rather than an end in itself.

DREXLER'S ARCHITECTURE

Drexler's assembler follows the von Neumann kinematic

architecture, but is specialized for dealing with systems made of atoms. The essential components in Drexler's assembler are shown in figure.

The emphasis here (in contrast toe von Neumann's proposal) is on small size. The computer and constructor both shrink to the molecular scale, while the constructor takes on additional detail consistent with the desire to manipulate molecular structures with atomic precision.

The molecular constructor has two major subsystems:

- A positional capability
- The tip chemistry.

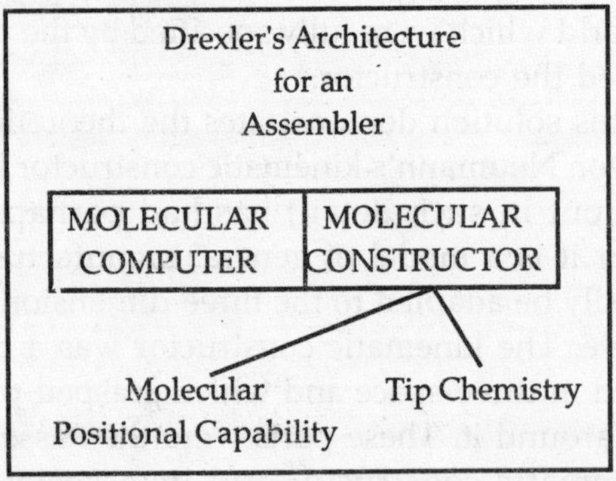

The positional capability might be provided by one or more small robotic arms, or alternatively might be provided by any one of a wide range of devices that provide positional control . The emphasis, though, is on a positional device that is very small in scale: perhaps 0.1 microns (100 nanometers) or so in size. The tip chemistry is logically similar to the ability of the von Neumann universal constructor to alter the state of a cell at the tip of the arm, but now the change in "state" corresponds to a change in molecular structure. That is, we must specify a set of well defined chemical reactions that take place at the tip of the arm, and this set must be sufficient to allow the synthesis of the structures of interest.

It is worth noting that current methods in computational chemistry are sufficient to model the kinds of structures that will appear in a broad class of molecular machines, including all of the structures and reactions needed for some assemblers

DEVICE SIZE

Drexler's proposal for molecular mechanical logic is the most compact and, from the system point of view, the best worked out. The logic elements ("locks," roughly the equivalent of a single transistor) need occupy a volume of only a few cubic nanometers.

Even including system overhead (power, connections, etc). the volume per element should still be less than 100 cubic nanometers. A 10,000 element logic system (enough to hold a small processor) would occupy a cube no more than 100 nanometers on a side.

That is, a volume only slightly larger than 0.001 cubic microns would be sufficient to hold a small computer. This compares favorably with the volume of a typical cell (thousands of cubic microns) and is even substantially smaller than subcellular organelles.

Operating continuously at a gigahertz such a computer would use less than 10^{-9} watts. By comparison, the human body uses about 100 watts at rest and more during exercise. Slower operation and the use of reversible logic would reduce power consumption, quite possibly dramatically. A variety of molecular sensors and actuators would also fit in such a volume.

A molecular "robotic arm" less than 100 nanometers long should be quite feasible, as well as molecular binding sites 10 nanometers in size or less.

By contrast, a single red blood cell is about 8 microns in diameter (over 80 times larger in linear dimensions than our 100 nanometer processor). Devices of the size range suggested above (~0.1 microns) would easily fit in the circulatory system and would even be able to enter individual cells.

APPLICATIONS

Given such molecular tools, we could design a small device able to identify and kill cancer cells. The device would have a small computer, several binding sites to determine the concentration of specific molecules, and a supply of some poison which could be selectively released and was able to kill a cell identified as cancerous. The device would circulate freely throughout the body, and would periodically sample its environment by determining whether the binding sites were or were not occupied.

Occupancy statistics would allow determination of concentration. Today's monoclonal antibodies are able to bind to only a single type of protein or other antigen, and have not proven effective against most cancers.

The cancer killing device suggested here could incorporate a dozen different binding sites and so could monitor the concentrations of a dozen different types of molecules. The computer could determine if the profile of concentrations fit a pre-programmed "cancerous" profile and would, when a cancerous profile was encountered, release the poison.

Beyond being able to determine the concentrations of different compounds, the cancer killer could also determine local pressure. A pressure sensor little more than 10 nanometers on a side would be sufficient to detect pressure changes of less than 0.1 atmospheres (a little over a pound per square inch.

See, for example, the discussion of *Nanosystems*for the kind of analysis involved. One atmosphere is ~10^5 Pascals, so PV in this case would be $(0.1 \times 10^5) \times (10^{-8})^3$ or $10^4 \times 10^{-24}$ or 10^{-20} joules. Multiple samples would be required to achieve reliable operation, as kT is 4×10^{-21} joules at body temperature.

Linear increases in sensor volume would produce exponential increases in immunity to thermal noise and linear improvements in pressure sensitivity if that were to prove useful. Doubling the linear dimensions of the sensor would produce an eight-fold increase in both volume and pressure sensitivity).

As acoustic signals in the megahertz range are commonly employed in diagnostics (ultrasound imaging of pregnant women, for example), the ability to detect such signals would permit the cancer killer to safely receive broadcast instructions. By using several macroscopic acoustic signal sources, the cancer killer could determine its location within the body much as a radio receiver on earth can use the transmissions from several satellites to determine its position (as in the widely used GPS system).

Megahertz transmission frequencies would also permit multiple samples of the pressure to be taken from the pressure sensor, as the CPU would be operating at gigahertz frequencies.

The cancer killer could thus determine that it was located in (say) the big toe. If the objective was to kill a colon cancer, the cancer killer in the big toe would not release its poison. Very precise control over location of the cancer killer's activities could thus be achieved.

The cancer killer could readily be reprogrammed to attack different targets (and could, in fact, be reprogrammed via acoustic signals transmitted while it was in the body). This general architecture could provide a flexible method of destroying unwanted structures (bacterial infestations, etc).

Providing Oxygen

A second application would be to provide metabolic support in the event of impaired circulation. Poor blood flow, caused by a variety of conditions, can result in serious tissue damage. A major cause of tissue damage is inadequate oxygen.

A simple method of improving the levels of available oxygen despite reduced blood flow would be to provide an "artificial red blood cell." We will consider a simple design here: a sphere with an internal diameter of 0.1 microns (100 nanometers) filled with high pressure oxygen at ~1,000 atmospheres (about 10^8 pascals).

The oxygen would be allowed to trickle out from the sphere at a constant rate (without feedback). Diamond has a Youngs modulus of about 10^{12} pascals. An atomically precise diamondoid structure should be able to tolerate a stress of greater than 5×10^{10} pascals (5% of the modulus).

Thus, a 0.1 micron sphere of oxygen at a pressure of 10^8 pascals could be contained by a hollow diamondoid sphere with an internal diameter of 0.1 microns and a thickness of less than one nanometer.

This thickness, thin as it is, results in an applied stress on the diamond of well under 1% of its modulus — from a purely structural point of view we should be able to use a very large "bucky ball," i.e., a sphere whose surface is a single layer of graphite.

Perhaps the most complex issue involved in the selection of the material is the reaction of the body's immune system. While some suitable surface structure should exist which does not trigger a response by the immune system — after all, there are many surfaces in the body that are not attacked — the selection of a specific surface structure will require further research.

To give a feeling for the range of possible surface structures, the hydrogenated diamond (111) surface could have a variety of "camouflauge" molecules covalently bound to its surface. A broad range of biological molecules could be anchored to the surface, either directly or via polymer tethers.

The Van der Waals' equation of state is $(p+a/v^2)(v-b) = RT$, where p is the pressure, v is the volume per mole, R is the universal gas constant, T is the temperature in Kelvins, and a and b are constants specific to the particular gas involved. For oxygen, a = 1.36 atm liter^2/mole^2 and b = 0.03186 liter/mole and R = 0.0820568 liter-atmospheres/mole-kelvin.

A mole of oxygen at 1,000 atmospheres and at body temperature (310 Kelvins) occupies 0.048 liters, or about 21 moles/liter. A mole of oxygen at 1 atmosphere and 310 Kelvins occupies 25.4 liters, or about 0.04 moles/liter.

This implies a compression of ~530 to 1. A resting human uses ~240 cc/minute of oxygen, so a liter of oxygen compressed to 1,000 atmospheres should be sufficient to maintain metabolism for about 36 hours (a day and a half).

It might be desirable to replace less than a liter of blood with our microspheres of compressed oxygen, but it should still be quite feasible to provide oxygen to tissue even when circulation is severely compromised for periods of at least many hours from a single infusion.

Controlled release of oxygen from the diamondoid sphere could be done using the selective transport method proposed by Drexler and illustrated in figure shows transport in the "wrong" direction (for this application), but simply reversing the direction of rotor motion would result in transport from inside the reservoir to the external fluid.

By driving a rotor at the right speed, oxygen could be released from the internal reservoir into the external environment at the desired rate.

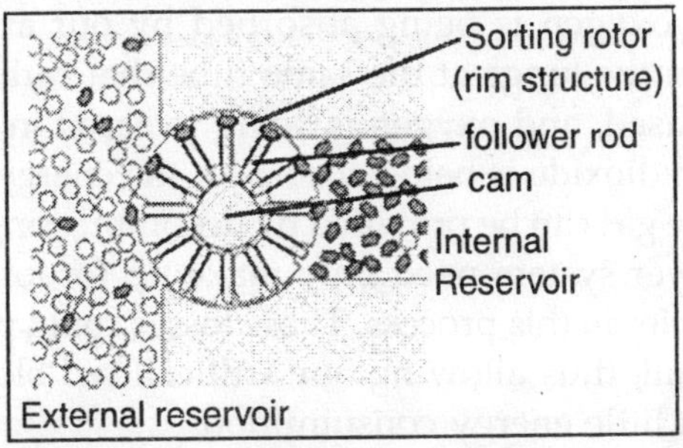

More sophisticated systems would release oxygen only when the measured external partial pressure of oxygen fell below a threshold level, and so could be used as an emergency reserve that would come into play only when normal circulation was (for some reason) interupted.

Full replacement of red blood cells would involve the design of devices able to absorb and compress oxygen when the partial pressure was above a high threshold (as in the lungs) while releasing it when the partial pressure was below a lower threshold (as in tissues using oxygen).

In this case, selective transport of oxygen into an internal reservoir (by, for example, the method shown in Figure would be required. If a single stage did not provide a sufficiently selective transport system, a multi-staged or cascaded system could be used.

Compression of oxygen would presumably require a power system, perhaps taking energy from the combustion of glucose and oxygen (thus permitting free operation in tissue).

Release of the compressed oxygen should allow recovery of a significant fraction of the energy used to compress it, so the total power consumed by such a device need not be great. If the device were to simultaneously absorb carbon dioxide when it was present at high concentrations (in the tissue) and release it when it was at low concentrations (in the lungs), then it would also provide a method of removing one of the major products of metabolic activity.

Calculations similar to those given above imply a human's oxygen intake and carbon dioxide output could both be handled for a period of about a day by about a liter of small spheres. As oxygen is being absorbed by our artificial red blood cells in the lungs at the same time that carbon dioxide is being released, and oxygen is being released in the tissues when carbon dioxide is being absorbed, the energy needed to compress one gas can be provided by decompressing the other.

The power system need only make up for losses caused by inefficiencies in this process. These losses could presumably be made small, thus allowing our artificial red blood cells to operate with little energy consumption.

By comparison, a liter of blood normally contains ~0.2 liters of oxygen, while one liter of our spheres contained ~530 liters of oxygen (where "liter of oxygen" means, as is common in the literature on human oxygen consumption, one liter of the gas under standard conditions of temperature and pressure).

Thus, our spheres are over 2,000 times more efficient per unit volume than blood; taking into account that blood is only about half occupied by red blood cells, our spheres are over 1,000 times more efficient than red blood cells.

Failure of a 0.1 micron sphere would result in creation of a bubble of oxygen less than 1 micron in diameter. Occasional failures could be tolerated. Given the extremely low defect rates projected for nanotechnology, such failures should be very infrequent.

Artificial Mitochondria

While providing oxygen to healthy tissue should maintain metabolism, tissues already suffering from ischemic injury (tissue injury caused by loss of blood flow) might no longer be able to properly metabolize oxygen. In particular, the mitochondria will, at some point, fail. Increased oxygen levels in the presence of nonfunctional or partially functional mitochondria will be ineffective in restoring the tissue. However, more direct metabolic support could be provided.

The direct release of ATP, coupled with selective release or absorption of critical metabolites (using the kind of selective transport system mentioned earlier), should be effective in restoring cellular function even when mitochondrial function had been compromised.

The devices restoring metabolite levels, injected into the body, should be able to operate autonomously for many hours (depending on power requirements, the storage capacity of the device and the release and uptake rates required to maintain metabolite levels).

While levels of critical metabolites could be restored, other damage caused during the ischemic event would also have to be dealt with. In particular, there might have been significant free radical damage to various molecular structures within the cell, including its DNA.

If damage was significant restoring metabolite levels would be insufficient, by itself, to restore the cell to a healthy state. Various options could be pursued at this point. If the cellular condition was deteriorating (unchecked by the normal homeostatic mechanisms, which presumably would cease to function when cellular energy levels fell below a critical value), some general method of slowing further deterioration would be desirable.

Cooling of the tissue, or the injection of compounds that would slow or block deteriorative reactions would be desirable. As autonomous molecular machines with externally provided power could be used to restore function, maintaining function in the tissue itself would no longer be critical. Deliberately turning off the metabolism of the cell to prevent further damage would become a feasible option.

Following some interval of reduced (or even absent) metabolic activity during which damage was repaired, tissue metabolism could be restarted again in a controlled fashion. It is clear that this approach should be able to reverse substantially greater damage than can be dealt with today.

A primary reason for this is that autonomous molecular machines using externally provided power would be able to continue operating even when the tissue itself was no longer functional. We would finally have an ability to heal injured cells, instead of simply helping injured cells to heal themselves.

MEDICAL RESEARCH

Advances in medical technology necessarily depend on our understanding of living systems. With the kind of devices discussed earlier, we should be able to explore and analyse living systems in greater detail than ever before considered possible. Autonomous molecular machines, operating in the human body, could monitor levels of different compounds and store that information in internal memory.

They could determine both their location and the time. Thus, information could be gathered about changing conditions inside the body, and that information could be tied to both the location and the time of collection. Physical samples of small volumes (nano tissue samples) could likewise be taken. These molecular machines could then be filtered out of the blood supply and the stored information (and samples) could be analysed. This would provide a picture of activities within healthy or injured tissue. This new knowledge would give us new insights and new approaches to curing the sick and healing the injured.

ENTIRE SYSTEM

More dramatically, it should be feasible to take "snapshots" of tissue samples and analyse the structure down to the molecular level. First, a small tissue sample could be either fixed or frozen. Chemical fixation can be used to rapidly block most tissue changes.

Ultra fast freezing of small tissue samples is an effective method of halting essentially all chemical processes and diffusion of all molecules. Once fixed or frozen, the tissue sample could be analysed in a leisurely fashion. With nanotechnology (and indeed, to some extent with current STM and AFM technologies, though rather more expensively) it should be feasible to scan the tissue surface in molecular detail, and store that information in a computer. Once the surface had been scanned, it could be removed in a very selective and precise fashion, and scanned again.

As an example, the use of a positionally controlled carbene has been proposed for use in the synthesis of complex diamondoid structures. Such a positionally controlled carbene is highly reactive and, if positioned at an appropriate site on the surface of the tissue being analysed, would readily react with a surface molecule.

This surface molecule could then be removed. A wide variety of other "sticky" molecular tools could be brought up to the surface and allowed to react with surface molecules, which could then be removed, exposing the layers beneath. The use of a positionally controlled carbene implies that the environment in which it is used must be inert.

This requirement could be satisfied by analyzing the tissue sample at very low temperature (a few Kelvins) and in a very good vacuum. Under these conditions the tissue specimen would remain stable during even a protracted analysis process.

While this process can readily be envisioned for very small structures, nanotechnology should make massive parallelism feasible. That is, a single positional device could be used at a certain speed to provide information about a certain (rather small) volume of tissue in a reasonable time.

Nanotechnology should permit the manufacture of a large number of small devices, each able to analyse a small volume. Given enough such devices operating in parallel, larger volumes could be analysed and the information from many individual devices integrated to provide a coherent picture of the larger whole. Effective use of this option will require massive computational power — which will also be made feasible with nanotechnology. Estimates of the computational power that should be provided by nanotechnology exceed 10^24 logic operations per second for a single desktop computer . This amount of raw computational power should make control of a large number of parallel devices feasible, and should permit integration and analysis of the information so obtained. In short, tissue samples could be "frozen" (either literally by ultrafast cooling or figuratively by chemical fixation) and the entire resulting tissue sample could be analysed down to the level of individual molecules.

The information so obtained could be processed by computers able to handle the flood of data produced. The resulting "snapshots" will provide us with an instantaneous look at metabolic and cellular activities across even relatively large volumes of tissue. Such an ability should revolutionize our understanding of the complex processes that take place in living systems.

The possibility of truly revolutionary advances in our medical abilities has also created renewed interest in cryonics. The abilities discussed here might well take years or decades to develop. It is quite natural to ask: "When might we see these systems actually used?" The scientifically correct answer is, of course, "We don't know."

AC ELECTROKINETICS: APPLICATIONS FOR NANOTECHNOLOGY

The phenomena of dielectrophoresis and electrorotation, collectively referred to as AC electrokinetics, have been used for many years to study, manipulate and separate particles on the cellular (1 mm or more) scale. However, the technique has much to offer the expanding field of nanotechnology, that is the precise manipulation of particles on the nanometre scale. The principles of AC electrokinetics for particle manipulation, review the current state of AC Electrokinetic techniques for the manipulation of particles on the nanometre scale, and consider how these principles may be applied to nanotechnology.

There are two principal approaches to manufacturing on the molecular scale: self-assembly of machines from basic chemical building blocks (the "bottom up" approach) and assembly by manipulating components with much larger devices (the "top down" approach).

Whilst the former is considered to be an ideal through which nanotechnology will ultimately be implemented, the latter approach more readily achievable using current technology. Examples of this include the manipulation of single xenon atoms on a silicon surface and the trapping of single 3nm colloidal particles from solution using electrostatic methods. Another group of methods with potential for use in top-down manufacturing — collectively referred to as AC electrokinetics — will be discussed.

AC Electrokinetic techniques such as dielectrophoresis and electrorotation have been utilised for many years for the manipulation, separation and analysis of cellular-scale particles. The phenomenon occurs due to the interaction of induced dipoles with electric fields, and can be used to exhibit a variety of motions including attraction, repulsion and rotation by changing the nature of the dynamic field. In many ways, these forces may be viewed as an electrostatic equivalent to optical tweezers and optical spanners in that they exert translational and rotational forces on a body due to the interaction between a body and an imposed field gradient.

Recent advances in semiconductor manufacturing technology have enabled researchers to develop electrodes for manipulating macromolecules as small as 9kDa using both attractive and repulsive AC electrostatic forces, and to concentrate 14nm beads from solution. Trapping of single particles such as viruses and 93nm-diameter latex spheres in contactless potential energy wells has also been demonstrated.

AC Electrokinetics offers advantages over scanning-probe methods of nanoparticle manipulation in that the equipment used is simple, cheap and has no moving parts, relying entirely on the electrostatic interactions between the particle and dynamic electric field. Furthermore, there is theoretical evidence that as manufacturing technology further improves, single particles considerably smaller than presently studied using AC Electrokinetics may be manipulated.

Ultimately, such a technology will have obvious applications for the manipulation of single molecules. Recent studies have shown that the trapping efficiency of planar electrode arrays is dependent on a number of factors including the magnitude and dimensions of the electric field and the radius of the particle to be trapped. In this paper we will consider the ways in which AC Electrokinetics can benefit nanotechnology, and the constraints on the technique due to factors such as Brownian motion, heating of the medium, and electrode dimensions as the electrode array is miniaturised to the nanometre scale.

DIELECTROPHORESIS

Consider a dielectric particle suspended in a spatially non-uniform electric field such as that shown in figure. The applied field induces a dipole in the particle; the interaction of the induced dipole with the electric field generates a force. Due to the presence of a field gradient, these forces are not equal and there is a net movement. If the particle is more conductive than the medium around it, the dipole aligns with the field and the force acts up the field gradient towards the region of highest electric field. If the particle is less polarisable than the medium, the dipole aligns against the field and the particle is repelled from regions of high electric field. The force is dependent on the induced dipole, and is unaffected by the direction of the electric field, responding only to the field gradient. Since the alignment of the field is irrelevant, this force can also be generated in AC fields which has the advantage of reducing any *electrophoretic* force to zero.

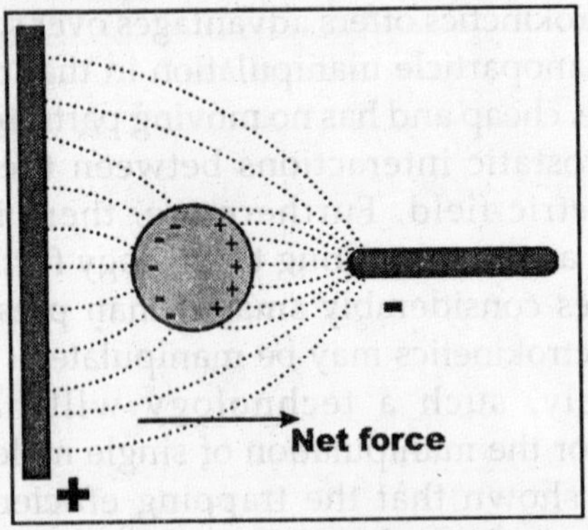

Fig: A schematic of a Polarisable Particle Suspended within a Point-Plane Electrode System.

When the particle polarises, the interaction between the dipolar charges with the local electric field produces a force. Due to the inhomogeneous nature of the electric field, the force is greater in the side facing the point than that on the side facing the plane, and there is net motion towards the point electrode. This effect is called positive dielectrophoresis. If the particle is less polarisable than the surrounding medium, the dipole will align counter to the field and the particle will be repelled from the high field regions, called negative dielectrophoresis.

This effect was first termed Dielectrophoresis by Pohl. The dielectrophoretic force, F_{DEP}, acting on a homogeneous, isotropic dielectric sphere, is given by:

$$F_{DEP} = 2\pi r^3 \varepsilon_m \operatorname{Re}[K(\omega)] \nabla E^2$$

where r is the particle radius, ε_m is the permittivity of the suspending medium, ∇ is the Del vector operator, E is the *rms.* electric field and ,

$$\operatorname{Re}[K(\omega)]$$

the real part of the Clausius-Mossotti factor, given by:

$$K(\omega) = \frac{\varepsilon_p^* - \varepsilon_m^*}{\varepsilon_p^* + 2\varepsilon_m^*}$$

where ε_m^* and ε_p^* are the complex permittivities of the medium and particle respectively, and

$$\varepsilon' = \varepsilon - \frac{j\sigma}{\omega}$$

with σ the conductivity, ε the permittivity and ω the angular frequency of the applied electric field. The frequency-dependence of,

$$Re[K(\omega)]$$

indicates that the force on the particle varies with the frequency. The magnitude of,

$$Re[K(\omega)]$$

also varies depending on whether the particle is more or less polarisable than the medium. If,

$$Re[K(\omega)]$$

is positive, then particles move to regions of highest field strength (positive dielectrophoresis); the converse is negative dielectrophoresis where particles are repelled from these regions. By careful construction of the electrode geometry which creates the electric field, it is possible to create electric field morphologies so that potential energy minima are bounded by regions of increasing electric field strengths.

In such electrodes, particles experiencing positive dielectrophoresis are attracted to the regions of highest electric field (typically the electrode edges, particularly where adjacent electrodes are close), whilst particles experiencing negative dielectrophoresis are trapped in isolated field minima.

ELECTROROTATION

If a polarisable particle is suspended in a rotating electric field, the induced dipole will form across the particle and rotate in synchrony with the field. However, if the angular velocity of the field is sufficiently large, the time taken for the dipole to form (the relaxation time of the dipole) becomes significant and the dipole will lag behind the field.

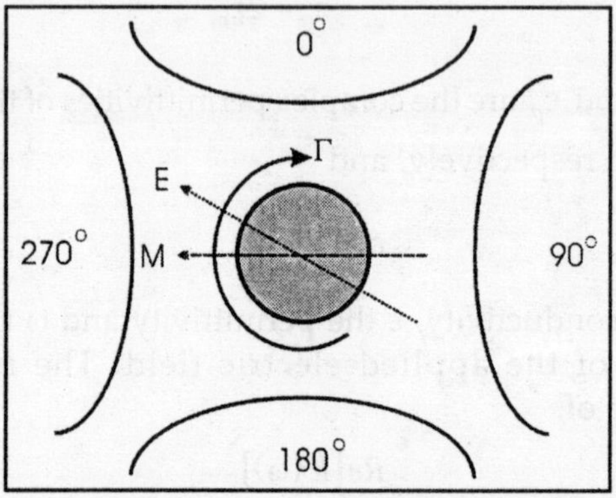

Fig: A schematic of a Polarisable Particle Suspended in a Rotating Electric Field Generated by four Electrodes with 90° Advancing Phase.

If the electric field E rotates sufficiently quickly, the induced dipole M will lag behind the electric field by an angle related to the time taken for the dipole to form (the relaxation time). The interaction between the electric field and the lagging dipole induces a torque G in the particle, causing the particle to rotate. This effect is known as *electrorotation*.

The general equation for time-averaged torque G experienced by a spherical polarisable particle of radius r suspended in a rotating electric field E is given by:

$$\Gamma = -4\pi\varepsilon_m r^3\, Im[K(\omega)]E^2$$

where $Im[K(\omega)]$ represents the imaginary component of the Clausius-Mossotti factor. The minus sign indicates that the dipole moment lags the electric field. When viscous drag is accounted for, the rotation rate R(ω) of the particle is given by:

$$R(\omega) = -\frac{\varepsilon_m\, Im[K(\omega)]E^2}{2\eta}$$

Note that there are two significant differences between the expression for torque as shown in equation.

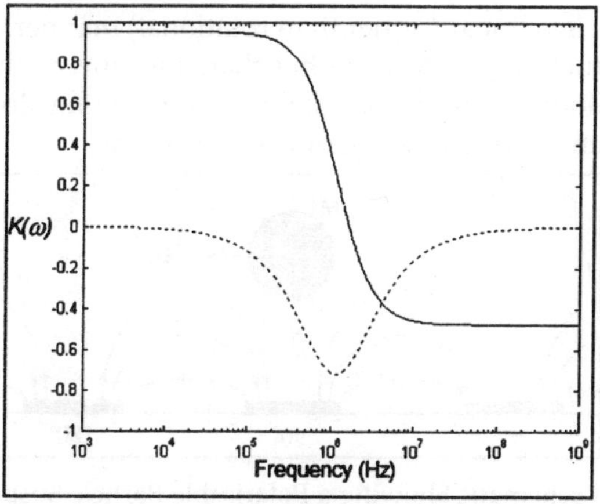

Fig. A Plot of the Real (Solid line) and Imaginary (Dotted line) Parts of the Clausius-Mossotti factor.

Firstly, the relationship with the electric field is as a function of the square of the electric field rather than of the *gradient* of the square of the electric field. Secondly, the torque depends on the imaginary rather than the real part of the Clausius-Mossotti factor. This is significant in that a particle may experience both dielectrophoresis and electrorotation simultaneously, and the magnitudes and directions of both are related to the interaction between the dielectric properties of particle and medium. As an example of this, a plot of the frequency-dependence of a solid particle exhibiting a single dielectric dispersion is shown in figure.

TRAVELLING-WAVE DIELECTROPHORESIS

Travelling-wave dielectrophoresis is a linear analogue of electrorotation; instead of electrodes being arranged in a circle, they are laid out as "tracks", but the relationship of phases (each electrode being 90° phase advanced from the last) remains. This produces an electric field wave which "travels" along the electrodes. When this wave interacts with a polarisable particle, a dipole is induced which moved with the electric field peak. If the travelling wave is moving sufficiently quickly, the dipole will lag behind the field in a similar manner to that observed in electrorotation. However, as the wave is travelling in a linear (rather than rotational) manner, the result is the induction of a force rather than a torque, with particles moving along the electrodes like a train travelling along tracks. This is demonstrated schematically in figure.

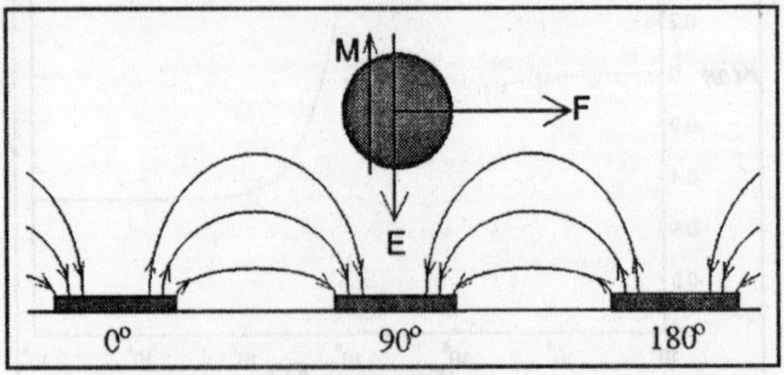

Fig. A schematic Showing a Polarisable Particle Suspended in a Travelling Electric Field Generated by Electrodes on which the Applied Potential is 90° Phase-advanced with Respect to the Electrode to its Left.

If the electric field E moves sufficiently quickly, the induced dipole M will lag behind the electric field, inducing a force F in the particle. This causes the particle to move along the electrodes, a phenomenon known as *travelling-wave dielectrophoresis*.

The value of the force F_{TWD} *is given by equation*:

$$F_{TWD} = \frac{-4\pi\varepsilon_m \gamma^3 Im[K(\omega)] E^2}{\gamma}$$

where l is the wavelength of the travelling wave. This phenomenon was first observed by Batchelder. However, it remained largely unexplored until work by Masuda *et al.* on using *synchronously* induced motion in low frequency travelling fields that travelling wave dielectrophoresis became significant. This work led to Fuhr *et al.* effectively rediscovering asynchronous travelling wave dielectrophoresis. Since then, a large corpus of study has been conducted on this technique, including theoretical studies and demonstrations of practical devices for electrostatic pumping and biotechnological applications.

ELECTRODE CONFIGURATIONS

In practice, dielectrophoretic manipulation of particles is performed on planar electrode arrays. Such electrode arrays employ patterned electrodes fabricated of gold or a similar conductor deposited on glass or silicon. A common electrode configuration used in AC Electrokinetics research is a quadrupole arrangement where four electrodes point towards a central enclosed region, as shown in the schematic in figure.

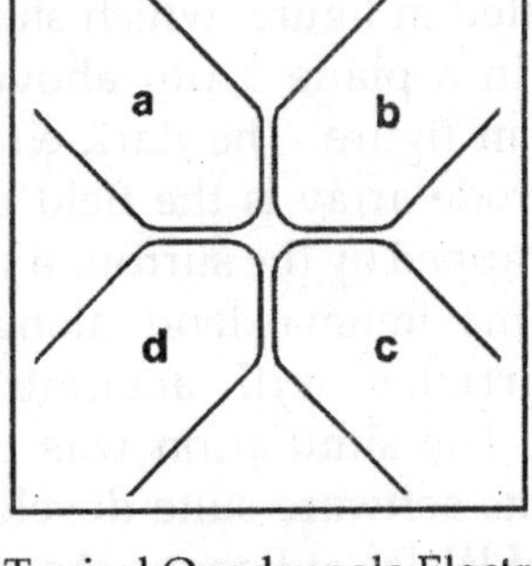

Fig: A Schematic of Typical Quadrupole Electrode Microstructures used in Dielectrophoresis Experiments.

The electrodes are typically fabricated of gold on a glass microscope slide. The gap between opposing electrodes in the centre of the array is typically of the order 10-50 μm across, but can be as small as 500nm or as large as 1mm.

To induce dielectrophoretic motion in particles suspended near the electrode array, electrodes would be energised such that a and c are of the same phase, and b and d are in antiphase to them. To cause electrorotation in particles within the central gap, electrodes b, c and d would be phase-shifted by 90°, 180° and 270° with respect to electrode a.

These electrodes may be of a variety of designs, according to the degree of field non-uniformity required. The advantage with this arrangement is twofold.

Firstly, electrodes can easily be used for dielectrophoresis, electrorotation or both. By energising the electrodes such that the phase angle of applied AC field between adjacent electrodes is 180°, particles will experience only dielectrophoresis. If the phase angle is reduced such that adjacent electrodes differ by 90°, the particle will experience both electrorotation and dielectrophoresis. The second advantage of the quadrupole design is that at the centre of the electrode array, there exists a well-defined electric field minimum, which is surrounded by well-defined field maxima at the electrode edges. This definition is advantageous when attempting to manipulate nanometre-sized particles, where Brownian motion is significant and may be sufficient to enable particles to "escape" from a weak forcefield "trap".

This is illustrated in figure, which shows a simulation of the electric field in a plane 5 μm above the plane of the electrodes shown in figure. The dark circular region at the centre of the electrode array is the field minimum in which particles become trapped by the surrounding ring of high field gradient. The field maximum extends along the inter-electrode gaps, where particles will accumulate by positive dielectrophoresis. The simulation was performed using a Method of Moments software suite developed by the author at the University of Wales at Bangor. As can be seen, there is an electric field minimum (dark region) at the centre of the electrode array, surrounded by a ring of high electric field gradient. Particles experiencing negative dielectrophoresis are repelled into this minimum and become trapped.

The electric field strength is high (white regions) along the electrode edges, where particles collect by positive dielectrophoresis.

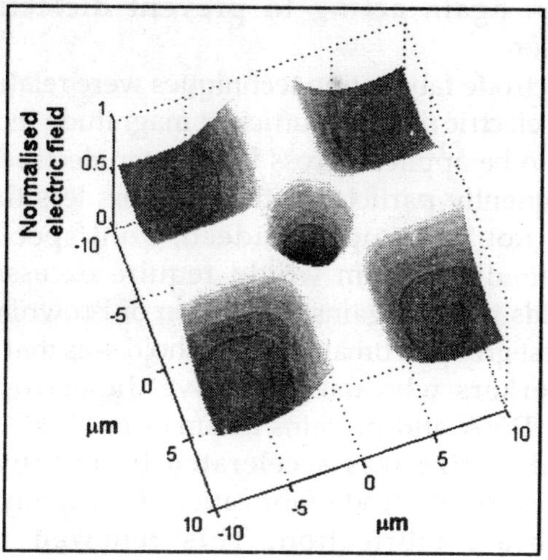

Fig: A Simulation of the Electric Field in the Plane 5 µm above the Electrode array.

Advances in fabrication technology such as the use of electron beam lithography have allowed the manufacture of electrodes with feature sizes of the order 100 nm, covering many square cm. By reducing the size of inter-electrode gaps, electric fields of 10^7 Vm^{-1} can be generated by applying 10V across a 1 µm inter-electrode gap. More advanced dielectrophoretic traps have been developed in which two sets of electrodes were stacked so as to form an octopole field "cage". This type of geometry has allowed the controlled manipulation of populations of particles in all 3 dimensions.

APPLICATIONS TO NANOTECHNOLOGY

Concentration of Colloids from Solution

Since the force experienced by a particle undergoing

dielectrophoresis scales as a function of the particle volume, it was believed for many years that a lower threshold of particle size existed, below which the dielectrophoretic force would be overcome by Brownian motion. It was held that to increase the force would require electric fields of such magnitude that local medium heating would increase local fluid flow, again acting to prevent dielectrophoretic manipulation.

As electrode fabrication techniques were relatively crude, generating electric fields of sufficient magnitude required large potentials to be applied across large inter-electrode volumes, and consequently particles with diameters less than about 1 μm could not be trapped. Indeed, Pohl speculated that particles smaller 500nm would require excessively large electric fields to trap against the action of Brownian motion.

The first group to break this threshold was that of Washizu and co-workers who used positive dielectrophoresis to precipitate DNA and proteins as small as 25 kDa. This step downward in size was accelerated by improvements in technologies for electrode fabrication, principally the use of electron beam fabrication. This renewed interest in manipulation of sub-micron particles, and subsequent work by Fuhr and co-workers and Green *et al.* demonstrated that viruses of 100 nm diameter could be manipulated using *negative* dielectrophoresis.

It was also demonstrated that latex spheres of 14 nm diameter could be trapped by either positive or negative dielectrophoresis, as demonstrated by Müller *et al.* and reproduced here in figure. Subsequent work by Hughes and co-workers demonstrated that by varying the frequency of the applied electric field, Herpes viruses can be trapped using either positive or negative electric fields. Another study demonstrated that molecules of the 68 kDa-protein avidin can be concentrated from solution by both positive and negative dielectrophoresis.

The electrodes were of similar construction to those shown in figure, with 500 nm between adjacent electrodes and 2 μm across the inter-electrode gap. In photograph (a), the spheres collect in the high-field regions between the electrode edges when a $8V_{pp}$, 2MHz signal is applied. In photograph (b) the spheres collect in a ball at the centre of the trap, which is levitated a few μm above the electrode plane when the applied signal is changed to $10V_{pp}$, 10MHz. Collection of particles is rapid and begins immediately after the field is applied. This experiment was performed at the Humboldt University of Berlin, reproduced with permission. Taken from Müller *et al.*

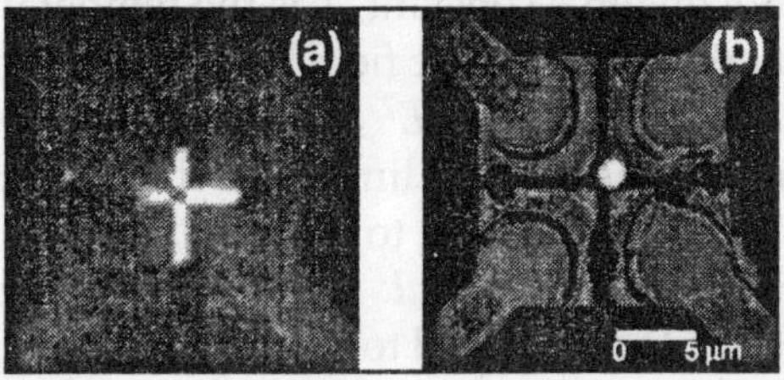

Fig: A Photograph of 14 nm Diameter, Fluorescently Labelled Latex Spheres Precipitated from an Aqueous Solution (Conductivity Adjusted to 2. 5 mSm^{-1} by Addition of KCl) by Dielectrophoresis.

The most obvious nanotechnological application of this technique is in the concentrating of "parts" of molecular machinery to one site. This could be either in the form of bringing together of two interlocking molecules in a highly dilute solution, or collecting "fuel" for a nanomachine from solution. Furthermore, by attracting microspheres and colloidal gold particles which have been functionalised using antibodies, Velev and Kaler have used dielectrophoresis to construct microscopic biosensors by "stacking" particles of different types, which bind on contact.

Manipulation of Single Nanoparticles

In 1996, Müller *et al.* Demonstrated that a single 1 μm latex sphere could be trapped by negative dielectrophoresis, and subsequently repeated the experiment with a latex sphere of 650 nm diameter. Further work by Hughes and Morgan demonstrated that single Herpes viruses, viral capsids and latex spheres as small as 93 nm diameter could be held in stable traps using negative dielectrophoresis, as shown in figure. This study also considered in some detail the mathematical conditions for a particle to be trapped by either positive or negative dielectrophoresis, concluding that for stable trapping, an approximate minimum particle radius *r* is given by the expression:

$$r > \sqrt[3]{\frac{10kT}{\pi\varepsilon_m \Delta d Re[K(\omega)]\nabla E^2}}$$

where k is Bolzmann's constant, T is the temperature, and $\Delta\, d$ is the thickness of the electric field barrier of magnitude, ∇E^2. Since the factors $\Delta\, d$ and ∇E^2 are related to the size of the electrode array, they place limits on the minimum size of particle an array can be used to manipulate.

Studies by Hughes *et al.* indicated that for electrode structures with 5Vp-p applied to electrodes with 6 μm between opposing electrodes and 2 μm between adjacent electrodes, the minimum stable trapping radius for single particle is approximately 30 nm.

This research indicated that the key variable governing $\Delta\, d$ is the size of the centre of the trap, which in that research was an area of approximately 5 μm radius centred on the middle of the electrodes.

Therefore it should theoretically be possible to reduce the whole array to this size; this would benefit the use of this technique in nanotechnology since the manipulator would be of the order of a few microns across.

Furthermore, since $\Delta\, d$ scales linearly with distance and ∇E^2 scales as the inverse cube of distance, equation indicates that reducing the size of the trap by a factor *x* whilst maintaining the same activating potentials will increase the efficiency of the trap by reducing the minimum trappable particle radius by a factor $x^{2/3}$.

An electrode structure similar to that shown in figure, with 6 μm between opposing electrode tips and 2 μm between adjacent electrodes. The medium conductivity was adjusted to $5mSm^{-1}$ by addition of EDTA, and the applied signal was $5V_{pp}$, 15 MHz. The bead occupies the space at the point of the "spike" at the centre of the electrode structure. The vertical dimension was extracted from light emission data from fluorescence micrographs. This experiment was performed at the University of Glasgow, reproduced with permission.

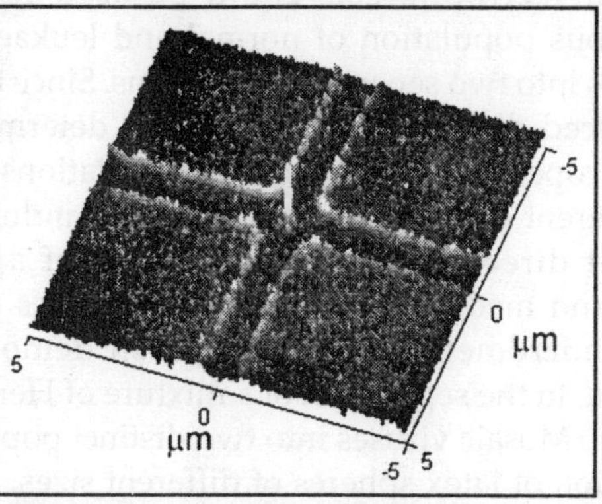

Fig: A pseudo-3D image of a Single 93nm Latex Sphere Suspended in a Trap by Negative Dielectrophoresis.

Due to the fact that this trapping is contactless, with particles being repelled from the electrode edges, particles may move upwards and will eventually leave the trap. In order to prevent this happening closed field cages, such as those fabricated by Schnelle *et al.* must be employed. An octopole cage of sufficiently small dimension could in theory allow full 3D manipulation of single particles of less than 30 nm diameter, and has been demonstrated to function for aggregates of 14 nm particles. The technology to produce such electrodes for single-molecule manipulation exists, but the principal problem to be overcome is the observation of the trapped particle.

An alternative approach would be to use an AFM tip to generate field non-uniformity, and use positive dielectrophoresis to attract particles to it. These could be positioned by the AFM and then "released" by changing the frequency of the signal to the AFM tip so as to repel the particles from it. The principle behind this — using positive dielectrophoresis to attract particles to a 3D, microscopic needle tip — has already been demonstrated to coat silicon tips with nanodiamond.

PARTICLE SEPARATION

In 1992 Gascoyne *et al.* demonstrated that positive and negative dielectrophoresis could be used to separate a heterogeneous population of normal and leukaemic murine erythrocytes into two separate populations. Since the direction of the induced dielectrophoretic force is determined by the dielectric properties of the particle, populations of particles having different dielectric properties can be induced to move in different directions by careful choice of applied field frequency and medium conductivity. Examples of this work in the submicrometre range have been demonstrated by Morgan *et al.* in the separation of a mixture of Herpes Simplex and Tobacco Mosaic viruses into two distinct populations and the separation of latex spheres of different sizes, or of similar size but different surface treatments. Another example of this is the separation of 93nm diameter latex spheres according to small variations in the surface charge of the spheres, as demonstrated by Green and Morgan.

In a nanotechnological context, the dielectrophoretic separation of particles allows for the manufacture of a tuneable electrostatic filter. By altering the frequency of the applied electric field, particles for two populations can be either both attracted to the electrode array, both repelled from it, or one can be attracted whilst the other is repelled. Since the effect is volumetric — repulsion acts for several microns into the solution, according to the magnitude of the electric field — the result is to alter the local concentration of the two particle types dramatically.

PARTICLE TRANSPORT

Whilst dielectrophoresis is useful for attracting particles to specific points, there are applications which require the moving of particles between one position and another. This requirement may be to relocate particles for processing, for separation purposes, or to gather in particles from a larger volume than can be achieved by conventional dielectrophoresis alone. Two means of particle transport have been demonstrated in the literature.

Firstly, conventional dielectrophoresis has been used to move 100 nm latex spheres in a thermal ratchet, where dielectrophoresis is used to rectify the motion of particles induced by Brownian motion. The induced motion can be either towards or against the ratchet potential, according to whether particles experience positive or negative dielectrophoresis, or the time between activation of ratcheting pulses. This technique has the advantage that since only two phases are required, the electrodes for inducing such motion are easy to fabricate.

The alternative, and more popular, means of transporting particles using AC electrokinetics is by the application of travelling-wave dielectrophoresis. Whilst the majority of work on travelling-wave dielectrophoresis has concerned micrometre-sized objects such as blood cells, some work has been performed on the concentration of particles on a surface using so-called "meander" electrodes. These structures use four electrodes in a series of interlocking spirals to generate a travelling wave; at the centre of the spiral, the electrodes form a quadrupole-type electrode array.

It has been demonstrated that by careful manipulation of the amplitudes of the potentials on these electrode structures, it is possible to "steer" the motion of particles across the array. Tools such as these could be used as the basis for "conveyor belts" for *factories on a chip,* wherein different chemical processes may be carried out of the same chip, with operations performed by electrostatic or chemical means and the resultant output transferred to a new process by AC Electrokinetic means.

ELECTROROTATION

Electrorotation was first examined scientifically in the early 1960s but was most fully explored by Arnold and Zimmermann, who established the technique for direct measurement of the value of *K(w)* by measuring the time taken for a particle to rotate. This application has little bearing to nanotechnology, not least due to the difficulty of seeing such small particles rotate, but electrorotation may be useful as a means of inducing rotation in molecular machinery. Recent work by Berry and Berg has involved using electrorotation to drive the molecular motor of *E. Coli* bacteria backwards at speeds of up to 2,000 Hz.

Since the electrodes involved in that work were applying 10 V across inter-electrode gaps of 50 μm or more, considerably higher fields (and hence induced torques) could be applied with electrodes with inter-electrode gaps of the order of 1 μm. If one considers that electrorotation is an electrostatic equivalent of laser spanners, then there exists a number of applications within current nanotechnological thought to which electrorotation might be applied. There are a number of advantages that favour electrorotation over its optical equivalent; most notable advantages are that electrorotation-induced torque is easily controlled by altering the frequency of the electric field, and that there does not need to be a direct optical path to the part to be manipulated. Electrorotation has been demonstrated for the induction of torque in micron-scale motors by Hagedorn et al.

Chapter 6

Types of Mechanical Reversible Logic

In real materials, energy dissipation occurs during successive cycles of compression and decompression of the material much as energy dissipation occurs when a gas is compressed by a piston. However, at sufficiently low temperatures, this mechanism of energy dissipation might no longer be operative. It is not impossible that energy dissipation can be made to fall exponentially as both temperature and speed of operation are reduced.

REVERSIBLE LOGIC

In a conventional computer the logical state of the system at time t (denoted S_t) will uniquely determine the state of the system at time t+1. If the function F maps the system state onto the successor state, then $S_{t+1} = F(S_t)$.

The requirement that the successor of a state be unique can be expressed as: if $F(S_t) \neq F(S'_t)$, then $S_t \neq S'_t$. In a logically reversible system, the state of the system at time t will uniquely determine the state of the system at both time t–1 and t+1.

This can be expressed as: if $F(S_t) \neq F(S'_t)$, then $S_t \neq S'_t$ *and* if $F^{-1}(S_t) \neq F^{-1}(S'_t)$, then $S_t \neq S'_t$. By contrast, in an irreversible system it will sometimes be the case that $F^{-1}(S_t) \neq F^{-1}(S'_t)$ and $S_t = S'_t$. That is, a given state might have two (or more) distinct and different predecessor states.

This neglects any special handling that might be required for the initial and final states of the computation. The initial logical state might not have a predecessor, and the final logical state might not have a successor. Alternatively, the predecessor of the initial state might be the final state, and the successor of the final state might be the initial state (i.e., the computation could be a giant loop). These states are rare in most computations.

The laws of physics are fully reversible at the microscopic scale, and so a physical system implementing a logically irreversible system is faced with a problem: when the logically irreversible system maps two logical states onto a single result state, the physical system must somehow map two physical states onto a single result state, as well.

This, of course, is impossible (for the underlying laws of physics are reversible, and hence the physical system is reversible), and so something has to give.

In today's computational systems, a single logical state will be represented not by a single physical state, but by any one of several possible physical states (which differ from each other in only trivial ways).

We can rephrase this by saying that each logical state of the system is represented by a certain volume of phase space, e.g., by a certain range of possible physical states.

When we perform a logically irreversible operation and merge logical states, we must either compress the representation in phase space or increase the volume of phase space that represents a logical state. If we ban indefinite growth in phase space then we are forced (at some point) to compress the phase space representation of the logical state. Phase space, however, is incompressible (which corresponds to the statement that merging two physical states is impossible, for the laws of physics are reversible).

As a consequence, irreversible operations force an increase in the volume of phase space occupied by the system. However, if the allowed computational degrees of freedom in the system are not allowed to expand, then this increase in phase space must take place in the non-computational degrees of freedom, e.g., waste heat.

Because mapping two logical states onto a single output state effectively erases a bit of information, the fact that irreversible logic must dissipate heat can also be stated as: erasing information must dissipate heat. Erasing a single bit of information generates at least $\ln(2) \times kT$ joules of heat.

By contrast a logically reversible system can be mapped onto the reversible physical world without the mismatch that occurs with logically irreversible systems.

There is no need to compress phase space, and the mapping between logical and physical representations need not perform a basically non-physical act, e.g., the erasure of information. As a consequence, there is no fundamental need to dissipate heat during the course of a computation.

This discussion of reversible logic has been given in terms of a computing system, rather than in terms of individual logic elements. This avoids a problem which can cause confusion: one and the same logic element can be viewed as either reversible or as irreversible, depending on the system context in which it is used. For example, AND gates are often cited as "irreversible" logic devices.

As commonly implemented and used, AND gates are indeed irreversible. However, it is possible to use some types of AND gates in a reversible fashion. All we need to do is insure that the operation of the AND gate takes place in a system context which does not erase information.

If the two inputs to an AND gate are erased following computation of the result then we must dissipate heat. If, however, the two inputs are *not* erased, then we have *not* erased any information and the computation need not be irreversible nor fundamentally dissipative.

This is a general principle: any irreversible combinational function F can be embedded in a larger reversible function F′. We might define F′ as: $F'(x) = \langle x, F(x) \rangle$. The new function F′ simply computes F and then concatenate the input x onto the output, insuring that we have not erased information.

We could, if convenient, retain even more information. If, during the course of computing F, we happen to compute various intermediate results $i_0, i_1 \ldots i_n$, then we could retain these intermediate results as well. We could define $F'' = \langle x, i_0, i_1 \ldots i_n, F(x) \rangle$.

F″ is also reversible, and by keeping the additional information its implementation might be simplified. As a consequence it is possible (in the proper context) to use AND and OR operations in a reversible computation.

The physical instantiation of such a logical computation can be done in a way which preserves local reversibility and is asymptotically nondissipative. Later in this paper we will sometimes compute irreversible functions F in a manner which is logically and thermodynamically reversible.

The actual implementation is done by retaining the intermediate values of the computation, as well as the input. This insures logical reversibility. In actual fact, the computation of F has been embedded in the computation of F″, where F″ is logically reversible.

In summary, reversible computations are consistent with the basic laws of physics at a microscopic scale, while irreversible computations are in some sense fundamentally incompatible.

The price we must pay for this incompatibility is heat. If we don't want to pay the price then we must learn to compute in harmony with the natural laws of physics, e.g., we must learn how to design reversible computers.

MECHANICAL COMPUTATION

The earliest example of a mechanical computer is Babbage's Analytical Engine. This device, had it been built, would have carried out the functions of a rather conventional computer but would have done so mechanically.

Remarkably, it was designed in the early 1800's. A working model of the simpler Difference Engine No. 2 was recently built by the British Museum of Science from Babbage's original blueprints using parts that would have been available in the 1800's, both in terms of the materials used and the precision of the milling.

A more recent reason for interest in mechanical logic has been that several proposals for reversible computation were basically mechanical in nature. Two mechanical designs that are well known in the reversible logic community are Fredkins "billiard ball" model of computation and Bennett's "Brownian Clockwork Turing Machine". Drexler's "rod logic" also implements combinational logic in a reversible manner. In the billiard ball model, billiard balls are fired into the computer and their ballistic trajectory — they bounce off each other and off fixed "mirrors" - defines an arbitrary reversible computation.

This is an example of a computation which dissipates as little energy as desired while rapidly computing an answer. In the clockwork Turing machine, the state of the computation can "drift" either forwards or backwards, subject to a small "driving force." The energy dissipation can be made arbitrarily small by making the driving force smaller and smaller, though at the cost of reducing the computational speed. More recently there has been interest in mechanical logic devices because:

- It should be possible to scale them to the molecular size range,
- At such sizes the speed of operation becomes reasonable (sub nanosecond switching times)
- Analyzing molecular mechanical logic devices is relatively simple (an important point when the devices can't yet be built).

Further, it should be possible to make molecular mechanical devices that are much smaller than molecular electronic devices (though the ultimate speed of operation of molecular mechanical devices will almost certainly be slower than that of molecular electronic devices).

ENERGY DISSIPATION

Whether mechanical, electronic, electromechanical, or whatever; the trends in energy dissipation will drive computer hardware ever more strongly towards the use of reversible computation. For the last 50 years the energy dissipation per gate operation has been declining with remarkable regularity.

Extrapolation of this trend shows the energy dissipation per device operation reaching kT (where k is Boltzmann's constant and T is the temperature in Kelvin) by the year 2015. (This assumes that T is 300 Kelvin - more on this later).

To gain some perspective on this consider that an "AND"

gate which has a power supply of one volt and which allows a single electron to go from that one volt supply to ground during the course of a switching operation, will dissipate one electron volt.

Although one electron volt is about forty times *kT* (and well above the theoretical limit), it will be difficult for simple improvements of current devices to reach even this level of energy dissipation. Extrapolating present trends, we should reach forty *kT* between the year 2000 and 2010, e.g., within ten to twenty years.

As a consequence, we can state quite confidently that one of three things will occur:

- The historic rate of decreasing energy dissipation per device operation will slow or halt in the next one or two decades
- We will operate computers at lower temperatures or
- We will develop new, novel, reversible computer systems that can beat the *kT* barrier.

The first option is quite unattractive. The heroic cooling methods used in the Cray 3 supercomputer to remove the heat generated by the computer's operation suggest that failure to reduce energy dissipation per gate operation would be a major limiting factor in future computer performance.

Further, although the raw cost of electrical power is not yet a major limitation, it would become so in the future if reductions in energy dissipation did not keep pace with advances in other areas.

The Wall Street Journal said "Computer systems currently account for 5% of commercial electricity use in the U.S., with the potential to grow to 10% by the year 2000." The second option will not reduce overall energy dissipation.

If we operated future devices at 3 Kelvin we could reduce energy dissipation by a factor of 100 - but for fundamental thermodynamic reasons the coefficient of performance of the refrigerator for the system can be at best 3K/(300K – 3K ~ 0.01[20]. Thus, the lower energy required per gate operation will be balanced by the increased energy needed by the refrigerator.

Further, in many applications low temperature operation is not an option. The use of liquid nitrogen in laptop computers or many embedded applications is not attractive. However, factors other than net energy savings can make low temperature operation worthwhile.

Some potentially attractive devices don't operate at higher temperatures: they have to be refrigerated or they don't work. A large mainframe computer made from such devices and operated at a low temperature might be the most economical method of delivering computational power in some cases. Refrigeration *per se*, however, does not seem too attractive.

Finally, and most attractively, we could develop reversible logic devices. Such devices would, in theory, allow energy dissipations indefinitely below kT per logic operation. While *some* barrier will likely eventually be encountered, the use of reversible logic should allow us to continue current trends in energy dissipation per logic operation for the longest possible time. Further research in this area is the most appropriate response to the rather limited range of possibilities that face us.

REVERSIBLE COMPUTER ARCHITECTURES

One problem is the need to develop novel "reversible" computer architectures. Such architectures have been discussed elsewhere. A wide variety of computations can be done in a reversible manner. Bennett concluded: "...for any $e > 0$, ordinary multitape Turing machines using time T and space S can be simulated by reversible ones using time $O(T^{1+e})$ and space $O(S \log T)$ or in linear time and space $O(ST^{e})$."

Even if we do not adopt new and novel computer architectures, simple applications of reversible computation could be made within the framework of existing architectures. A typical computer executes a sequence of instructions, and each instruction will typically change the contents of a single register or memory operation.

Although loading the result of the instruction into a register will normally be irreversible (it destroys the previous contents of the register) it is still the case that all *other* operations performed by the computer during instruction execution could in principle be made reversible. Thus, although we would have to dissipate roughly *kT* energy for each bit in the output register for each instruction execution, we need not use energy-dissipative logic devices throughout the computer.

We can go one step further without making any changes in computer architecture: the simple register-register add instruction $R_1 = R_1 + R_2$ is logically reversible. With proper hardware, this particular instruction could in principle be made to dissipate as little energy as we desired.

While irreversible instructions (e.g., $R_1 = 0$) would still dissipate greater energy, the overall energy dissipation of the computer could be reduced significantly if a reversible system were developed and used to implement the reversible instructions.

Of course, once the energy-wasting irreversible instructions were identified, compilers would learn to avoid using them. This would provide an entirely evolutionary path from the current irreversible designs to computer architectures that were as reversible as was practically feasible.

While there is debate about how far this trend can go it is clear that a significant percentage of computer operations can be made reversible - perhaps a remarkably high percentage.

SOME OTHER PROPOSALS

To be successful, we must develop appropriate fully reversible switching devices that are physically realizable. We shall use the term "physically realizable" to mean "implementable by an appropriate configuration of atoms."

Many of the earlier proposals do not appear to be physically realizable, although recent reversible proposals, particularly those based on conventional technologies such as CMOS or CCD's, are very clearly physically realizable.

An early electronic proposal by Likharev based on Josephson junctions "...is particularly significant, because it is a genuine example... of a system that has frictional forces proportional to velocity" according to Landauer. Likharev's "parametric quantron" is directly analogous to the second proposal made later in this paper. The parametric quantron is based on Josephson junctions and operates at low temperatures. Modern "high temperature" superconductors should allow operation at liquid nitrogen temperatures (although Josephson junctions have not yet been demonstrated to work at that temperature), but the low temperature of operation is still a significant disadvantage.

The parametric quantron uses magnetic fields to transmit logic information from device to device. It cannot be scaled to the size range considered here for molecular mechanical devices, and the speed-power trade-off in the parametric quantron might not be significantly better than the speed-power trade-off in molecular mechanical devices.

The first proposal discussed here is derived from the work of Drexler, who has argued persuasively that mechanical logic devices can be scaled to the molecular level. The scaling properties of current electronic devices are unlikely to be as good. Drexler's proposed "rod logic" is also reversible.

Drexler concluded that rod logic, when scaled to molecular size, should eventually be able to achieve an energy dissipation of somewhat less than kT at 300 Kelvins with gate delay times of 50 picoseconds.

Reducing the speed of operation would further reduce energy dissipation. Drexler, however, also argues that electronic devices will very likely prove superior in speed of operation, a conclusion supported by recent work by Merkle and Drexler.

The primary reason for pessimism about the speed of mechanical logic is that the nuclei of atoms have greater mass than electrons, and so any device which depends on movement of the nuclei seems doomed to be slower than a device that uses the motions of electrons.

The carbon nucleus is over 20,000 times as massive as an electron, and will normally accelerate roughly 20,000 times more slowly. An alternative view is that the speed of sound in solids is roughly 10^4 meters per second, whereas the speed of electronic signals is roughly 10^8 meters per second.

Speed is not the only important criterion for evaluating

logic devices. If small size is the deciding criterion then molecular mechanical reversible logic is more attractive. Small size might make it competitive for some applications (high density memory, for example) where speed is not paramount. While the greater mass of nuclei as compared with electrons results in slower speed, it also greatly reduces the distance through which the nuclei can tunnel.

One of the major limitations in scaling electronic devices to the smallest possible size is the ability of electrons to tunnel distances of a few nanometers. By contrast, devices whose function depends on the position of nuclei can be scaled to linear dimensions of two nanometers or less.

It might be that an electronic device only two nanometers across will be forever infeasible because a single electron can tunnel this distance (though such a claim must be viewed with caution). Thus, molecular mechanical devices might enjoy a fundamental advantage in terms of small size. Another limitation of mechanical devices is the sliding motion of one surface upon another.

This is not a significant problem if atomically perfect surfaces can be used as Drexler assumes, but current manufacturing techniques are not able to achieve this level of perfection. Drexler and others have argued that we will eventually be able to build most physically realizable structures, including complex molecular logic elements which are atomically precise.

While electromechanical logic devices employing sliding silicon parts have been built and are radiation resistant, the friction and wear caused when imperfect silicon surfaces slide over each other is a major drawback.

As we will demonstrate, a mechanical logic device need not involve a sliding motion of one surface past another. Elimination of the sliding motion results in several advantages. First, it simplifies the analysis of molecular-scale mechanical logic devices. Because such devices cannot be built with current technology, any claims about their performance must rest on theoretical analysis. By eliminating the sliding motion the analysis becomes simpler and more fool proof.

A second more practical advantage is the elimination of the sliding motion when larger devices are built with current technology. The use of non-sliding logic in silicon eliminates the wear and friction that the sliding action creates. The energy dissipation of the device can therefore be substantially reduced, and the operational lifetime extended.

BASIC IDEA

Drexler's rod logic is based on the observation that two objects cannot occupy the same place at the same time. If two objects both have parts that can occupy the same place when the objects move, then the motion of one object can be used to limit the motion of the other.

In rod logic the interacting objects are rods and the parts that interact are called "knobs." Normally, a rod has one degree of mechanical freedom: it can move back and forth along its own axis. Two rods that are set at right angles to each other can interact if they can be moved so that their "knobs" can occupy the same position.

This common region of space that both knobs can attempt to occupy is called the "lock." One rod can be moved so that its knob occupies the lock, but when this is attempted with the second rod, the knobs will collide and the motion of the second rod will be blocked.

The first knob to reach the lock is the "gate knob," and the second knob is the "probe knob." Using this terminology, if the gate knob occupies the lock, the probe knob is blocked and thus the motion of the second rod is also blocked.

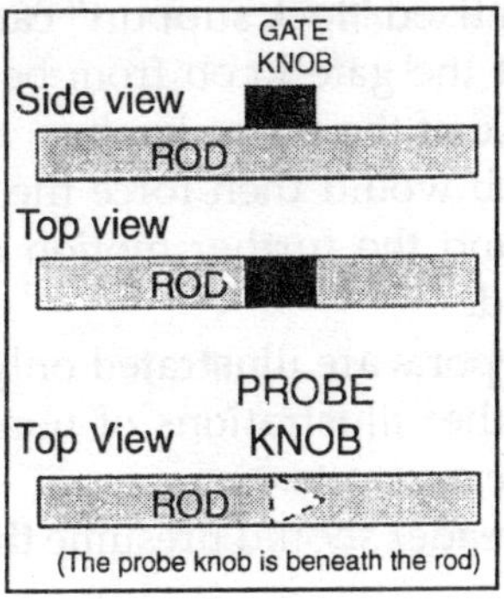

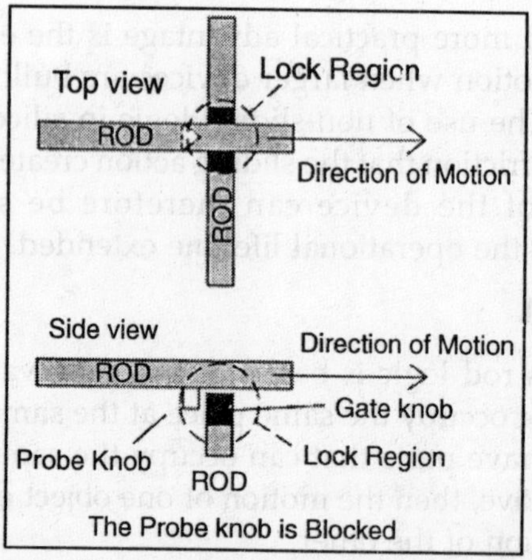

The rods must be constrained in their motions so that they cannot move except along the specified axis, nor rotate and change the location of the knob. Either of these illegal motions would let one rod slip past the other and cause a logic failure. Drexler constrained the rods by embedding them in a "matrix," a solid block of material with "channels" in which the rods are obliged to slide.

We eliminate the matrix by the simple expedient of mounting the rods on flexible supports, which hold the rods in position. When a rod is pulled, the supports are elastically deformed (which allows rod motion).

If the supports prevent lateral motion of the second rod then the collision of the probe knob with the gate knob will block the further motion of the probe knob. If the supports do not sufficiently prevent lateral motion of the second rod during the collision, then a fixed "lock support" can be placed behind the lock to prevent the gate knob from being pushed out of position by the force of the probe knob.

The probe knob would then force the gate knob against the lock support, and the further motion of the probe knob would be prevented.

The elastic supports are illustrated only in figure. For the sake of clarity, further illustrations of non-sliding logic will not show the elastic supports.

However, the reader should presume that each rod is held

in position by an appropriate set of supports, even though they are not shown. The length of the rods must also be controlled. Stretching of the rods would change the positions of the gate knobs. Should the gate knobs move a sufficient distance from their correct location from this cause, even though they continued to move only in the channel, the probe knobs would slip past them.

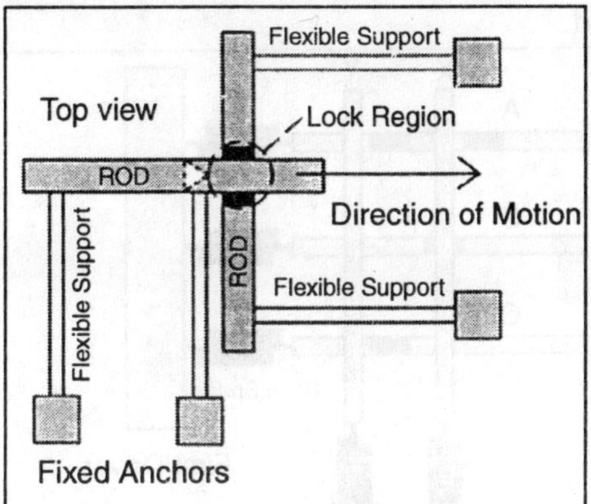

Rods Supported by Flexible Supports

While stretching is usually insignificant at the macroscopic scale, it is significant and must be taken into account on the molecular scale. Drexler reduced this source of error by applying tension to the ends of the rods.

When this is done the material from which the "rods" are made can be both flexible and compressible, and might be better described as string. The term "rod" should thus be viewed with some caution, as many rather un-rod-like structures can be used quite effectively. Rod logic is somewhat different from many popular logic systems. In most current systems, the output signal from a typical gate is available shortly after the input signals are applied. This does not occur in rod logic, which is fundamentally a clocked system.

Rods do not move unless pulled, and only when a rod is pulled can we tell if it is blocked or unblocked. The output signal is not available until after the input signals are applied and the output rod is pulled. For this reason, the temporal pattern used to pull the rods (the clocking sequence) is of critical importance in the operation of a rod-logic circuit. At each point in time a given rod can be:

- Pulled
- Released
- Left unchanged.

If it is pulled, it either moves or does not move, depending on whether or not the locks associated with the rod are blocked or unblocked.

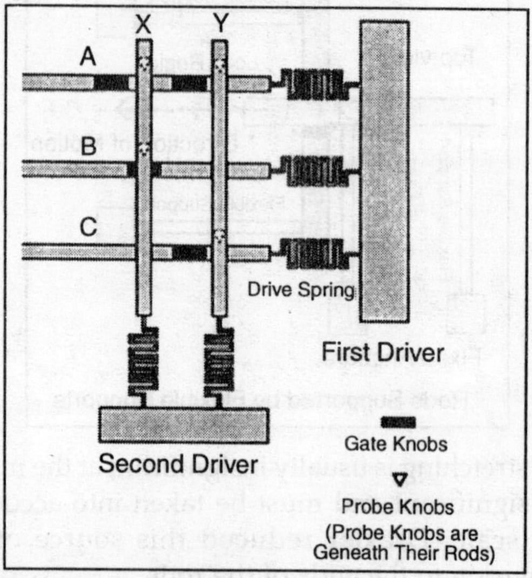

We now give an example of a complete cycle of operation for a simple logic function. Three inputs, A, B, and C, enter from the left along the three horizontal rods. The first driver is on the right, and three drive springs connect rods A, B, and C to the driver.

Two outputs, X and Y, exit downwards along the two vertical rods. Two drive springs connect the bottoms of the rod to the top of the second driver. At the intersection of each horizontal rod with each vertical rod, there can be a lock. Note that each lock can initially be either blocked (occupied by the gate knob) or unblocked (not occupied by the gate knob).

The cycle of operations proceeds as follows. First, the first driver moves to the right. This applies force to the three rods

A, B, and C through the three drive springs. Rods B and C are free to move, while rod A is blocked (by a lock not shown on the diagram). The drive spring for rod A extends, while the drive springs for rods B and C do not. The motion of rods B and C unblocks the gate at the intersection of B and X, but blocks the gate at the intersection of C and Y.

The two locks controlled by rod A continue to be unblocked. T. Second, the second driver moves downwards. This applies force to the two output rods X and Y. X is free to move, and does so, while Y is blocked. The drive spring for Y extends.

This is illustrated in Figure. Third, the second driver is released and rods X and Y resume their original positions. Fourth, the first driver is released and rods A, B and C resume their original positions.

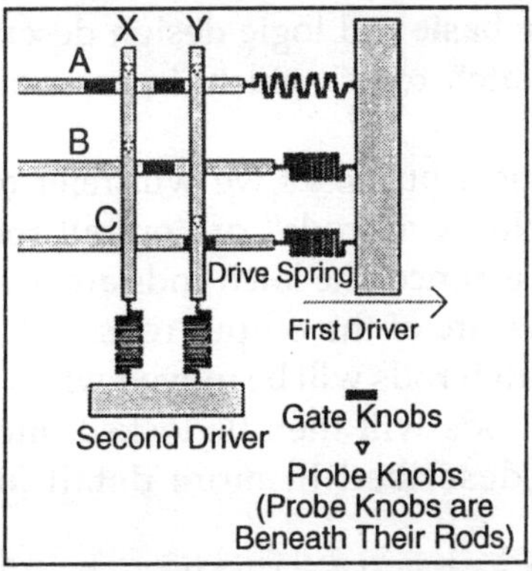

Note that cycles 1 and 2 compute an output, while cycles 3 and 4 reverse the motions involved in cycles 1 and 2, thereby "uncomputing" the output. Therefore this clocked sequence of actions is entirely reversible.

The energy dissipation per operation could, in principle, be made less than kT. If we attempted to change this sequence, e.g., we interchanged cycles 3 and 4 and released driver 1 before releasing driver 2, then the probe knob and gate knob would slide past each other at the lock at the intersection of C and Y. That is, the wrong sequence of clock signals would introduce sliding motion.

The clock sequence in rod logic must be carefully specified to prevent this kind of problem. In general, if we have an N level logic circuit, then we will need to divide the rods into N groups numbered from 1 to N. Each rod in a group will be connected to the corresponding driver. The drivers will be moved in sequence, starting with driver 1 and terminating with driver N.

The rods in group i will control the motion of the rods in group i+1. After the Nth level rods have been pulled, the drivers will be released in reverse sequence, starting with driver N and terminating with driver 1.

What has been described so far will allow the construction of an N level reversible combinational logic circuit. To be useful, we must also specify a latch which can store the output of the combinational circuit. This can be done relatively easily by modifying the basic rod logic design described so far. We define a set of "latch rods" which, in essence, are the N+1st logic level rods.

For pedagogical purposes we will refer to the Nth level rods as either "Nth level rods" or "output rods," whichever seems clearer. In essence, the latch rods are used to "latch up" or remember the state of the output rods.

The proper latch rods will be moved when the output rods move. The latch rods will then be locked into position. This arrangement is described in more detail in the next few paragraphs.

Each output rod will be lengthened and a new "drive knob" will be added near the end away from the drive spring. A "drive knob" is a knob whose purpose is to pull a "driven knob," and in consequence to move the rod to which the driven knob is attached, also called the "driven rod."

The driven rod, like all rods, will move along its axis. As a consequence, the driver rod and the driven rod must be parallel, for the motion of the drive rod will be transferred directly to the driven rod. The action of the driver knob is unlike the action of either a "probe knob," which is used to determine whether or not a lock is blocked, or a "gate knob," which is used to block and unblock a lock.

The drive knob on the output rod will be used to pull on the driven knob which is attached to the latch rod. In this case, the output rod is the driver rod and the latch rod is the driven rod. The latch rod is parallel to the output rod, not at right angles.

If the output rod is pulled and moves, then the driver knob on the output rod will pull on the driven knob on the latch rod, and the latch rod will also be pulled. Because the latch rod will not be blocked, it will also move.

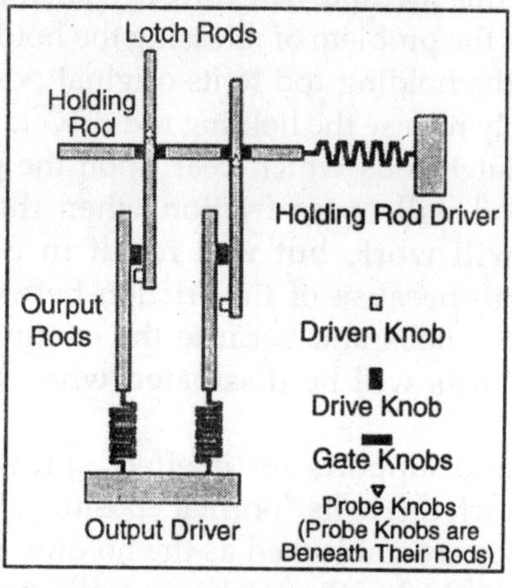

Once the latch rods have been moved into their proper positions, they are held in place by the "holding rod." The holding rod is at right angles to the latch rods, and is always free to move. The holding rod has gate knobs at all positions. All the locks between the latch rods and the holding rod are initially unblocked.

After the latch rods are in their correct position, the holding rod moves and all the gate knobs on the holding rod block all the gates. In those latch rods which have not moved, the probe knob on the latch rod is clear of the lock and has not moved through it.

On those latch rods which have moved, the probe knob has been moved through the lock and is also clear of it. Thus, in either case, the holding rod is free to move and the gate knobs on the holding rod can occupy all the corresponding locks. Once the locks are blocked, the output driver is released. The output rods move back to their original positions. Those latch rods that were moved are now held in place by the gate knobs on the holding rod and cannot return to their original positions.

The output rods are free to move at this point in the cycle, whether or not the latch rods have been blocked, because the motion of the driver knobs on the output rods is now away from the driven knobs.

Although this method will correctly latch up the output, it now presents the problem of releasing the holding rod driver and restoring the holding rod to its original position.

If we simply release the holding rod driver, then the probe knobs on the latch rods which bear upon the gate knobs on the holding rod will cause friction when the holding rod moves. This will work, but will result in greater energy dissipation both because of the friction between the probe knobs and gate knobs and because the energy stored in the latch rod supports will be dissipated when the latch rods spring free.

The latch rod supports are, in effect, springs that attempt to restore the latch rod to its "normal" position. When the latch rod probe knobs are unblocked as the holding rod gate knobs move out of the locks, the latch rod will snap back to this position and oscillate, thus dissipating energy.

This is in keeping with the irreversible nature of the operation: the information held in the latch rods will be destroyed, which requires dissipation of about kT per latch rod. A more clever method would be to "unwrite" the information held in the latch rods.

This differs fundamentally from erasing the information. In this technique, the information held in the latch rods is recomputed (by any method we might find convenient) and the output rods are driven with this freshly recomputed information. This recomputed information is identical to the actual contents of the latch.

In this way, the pressure on the holding rod gate knobs will be released, and the holding rod will again be free to move.

The holding rod driver can then be released, and the output driver (and all preceding drivers) can be released as usual. The information in the latch rods is "unwritten" in a reversible manner (no step in the unwriting process destroys information) and the energy dissipated per latch rod operation can in principle be less than *kT* during this process.

This allows us to remember the value computed by a combinational logic circuit in a set of latches. With some additional mechanism, we can compute an iterated reversible computation. This will allow us to perform the more or less normal sequence of operations in a computer.

That is, the state of a computation in a normal CPU is held in a series of latches (flip-flops) and that state is used as the input to a combinational logic circuit which computes the next state. The next state is then clocked into the flip-flops while the previous state is destroyed.

This sequence allows a conventional computer to execute a series of instructions by repeatedly computing the next state from the current state, and repeatedly replacing the current state with the next state.

Of course, in a reversible computation, we not only need to compute the next state from the current state, we must also be able to uniquely determine the current state from the next state, e.g., the computation must be reversible.

There are many ways of designing such a reversible computer. In the following paragraphs we describe a simple architecture for carrying out a reversible computation. We assume that we have a reversible combinational function whose iteration we feel is computationally useful (e.g., the function implements a single instruction or step of a reversible computation).

We shall call this function F. We start with the initial state of the computation stored in a set of initial latch rods which are held in position by a driven holding rod. We then compute F(initial latch rods), and store the result in a set of output latch rods, which are assumed to be initially empty. Having once computed the next state and stored it in the output latch rods, we must unwrite the initial latch rods. To do this, we compute

F^{-1}(output latch rods), which should be identical to the state already stored in the initial latch rods. Thus, not only do we have combinational logic for computing F, we have combinational logic for computing F^{-1}. This additional logic is absolutely necessary.

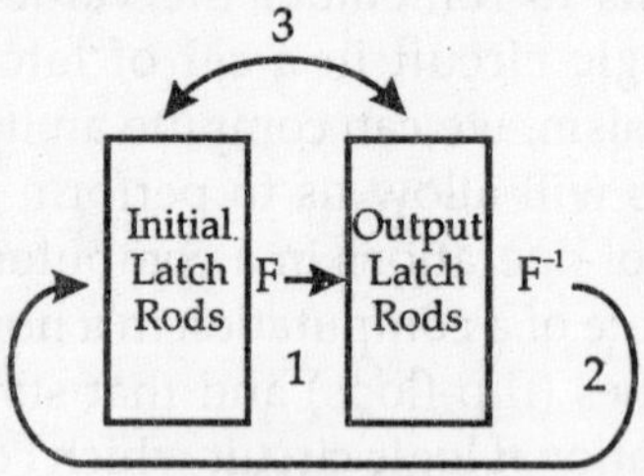

Flow of information while iterating a reversible function F

The summary of one full cycle of computation is illustrated in figure. In step 1, we compute output latch rods = F(initial latch rods). The output latch rods are left holding the output, while we uncompute the intermediate stages in the computation of F.

This leaves the input in the initial latch rods, and the output in the output latch rods. In step 2 we compute F^{-1}(output latch rods) and use this information to unwrite the initial latch rods. After unwriting the initial latch rods they hold 0. We then uncompute the intermediate stages in the computation of F^{-1}(output latch rods).

This leaves us in a state in which the output latch rods hold the next stage of the computation, while the input latch rods hold 0. In step 3 we exchange the contents of the output latch rods and the input latch rods. This exchange is actually done in two steps: first, we write the initial latch rods using the output latch rods as the source of information.

This operation can be done reversibly because we know the initial latch rods hold 0. Second, we unwrite the output latch rods using the initial latch rods as the source of information for the unwriting operation. After completion of these operations, we are right back where we started except that the initial latch rods hold F(contents of initial latch rods from start of previous cycle). All operations were done reversibly, and so can be carried out in an asymptotically nondissipative fashion.

ERRORS

Molecular mechanical systems can suffer failures because of thermal noise. These are not significant at the macroscopic level because larger structures are less affected by thermal noise than smaller structures.

If a molecular mechanical system has too high a thermal error rate, then the system can be increased in size until the problem is effectively controlled. Alternatively, the temperature can be reduced until thermal noise is insufficient to cause problems.

Even in molecular mechanical systems, however, thermal noise at room temperature can be effectively dealt with by appropriate design. In general, the parts in a molecular mechanical system will be held in place by some restoring force.

This restoring force will act like a spring and will tend to prevent the part from moving away from its assigned position. If the restoring force is linear in the displacement (which is often a good approximation) then we have:

$$f = k_s x$$

where k_s is the spring constant, x is the displacement, and f is the restoring force. It requires a certain amount of energy to displace the part, so we also have:

$$E(x) = k_s x^2/2$$

where E(x) is the energy of the system when the part is displaced by a distance x. In thermal equilibrium the probability of being displaced by an amount x falls off exponentially as the energy E(x) increases. The probability of being in error falls off even more rapidly in the displacement:

$$P(x) = \mathrm{sqrt}(k_s/2\ \mathrm{pi}\ kT)\ \exp((-x^2 k_s)/(2\ kT))$$

where P(x) is the probability of being displaced by a distance x from the correct position. The spring constant for a single carbon-carbon bond is approximately 440 newtons/

meter . At room temperature a spring constant of this value implies that a displacement of 0.05 nanometers has a probability of about 10^{-47}.

Even for a spring constant of 1 newton/meter, the probability of being displaced by a nanometer is about 10^{-43}. Molecular mechanical systems designed to tolerate positional errors on the order of several tenths of a nanometer can have very high reliability in the face of thermal noise at room temperature.

ENERGY LOSSES AT THE CONTACT SURFACES

When two parts contact each other, there will presumably be some loss of energy. If actual contact between two surfaces occurs, then a number of surface effects might take place. Bond-making and bond- breaking are likely to occur when the parts are large and the contact pressure is too high.

As the size and speed of the parts is reduced the contact pressure will at some point become small enough that it doesn't break individual bonds. When contact surfaces are deliberately designed to be stable during contact and the design deliberately keeps contact pressure low, bond breakage will not be a significant factor.

In this case, intermolecular forces between the two surfaces will become the dominant factor in any energy loss mechanism. Intermolecular forces can be described by an appropriate potential energy function and arise from the electric forces between the nuclei and the electrons (other forces are small enough that they can be neglected).

"Long range" intermolecular forces (acting over distances measured in angstroms to nanometers) can be conveniently divided into three categories: electrostatic, induction, and dispersion. The electrostatic contribution arises from the uneven distribution of charge on a molecule. Two molecules with uneven charge distribution (e.g., two dipoles) will experience a force caused by the attraction and repulsion of their charged regions. Induction forces arise when a molecule with a dipole moment induces a dipole in an otherwise non-polar molecule. Dispersion forces arise because fluctuations in the electron cloud around one molecule create an instantaneous dipole, which can then induce an instantaneous dipole in an adjacent molecule. This creates a net attractive force between two otherwise neutral molecules.

The interaction energy in the latter two cases falls off as the sixth power of distance, while the dipole-dipole interaction energy falls off as the third power of distance. Appropriate selection of materials and speed of operation can be used to minimize the energy loss during surface-surface contacts.

Drexler showed that energy losses from this source can be quite small (less than kT) even when switching speeds are 50 picoseconds.As an aside, even sliding motion need not cause problems. Provided that two sliding surfaces are structurally stable (so that the formation and breaking of chemical bonds is not an issue), the surface-surface interaction in this case will also be described by a potential energy function.

In the case of two surfaces that are constrained to move past each other in one dimension, the potential energy function is a simple function of a single variable. If the two surfaces are periodic (the atoms on the surfaces are regularly spaced) and if the periods for the two surfaces are the same (the two surfaces are made of the same material) then the potential energy as a function of position will have high peaks and low valleys: the two surfaces will resist sliding past one another.

If, however, the two surfaces are selected to have *different* periodicities (they are made of different materials, or the surfaces are skewed with respect to each other along the line of motion) then the potential energy as a function of position can be essentially flat. In this case, there will be little resistance to motion. If the motion of one surface past another is relatively slow then energy dissipation can be quite small. Thus, while sliding motion introduces an additional design problem, it is a problem that can be overcome if the surfaces can be built to atomic specifications.

ELIMINATING THE CONTACT SURFACES

It is possible to define a kind of logic in which there is

only a single large "part" with a complex shape. The state of the computation is encoded in the deformations of this single elastic block of material. The computation moves forwards as this block goes through periodic oscillatory motions. This eliminates not only sliding motions, but any contact at all between surfaces. To illustrate the principle underlying this kind of logic, consider the behaviour of an isolated elastic vertical beam when a compressive force is applied from the top. Imagine, for example, that you hold a business card vertically in your hand by its edges, forefinger on top and thumb on the bottom.

When you squeeze the card it will buckle in either of two directions. If the card was initially straight, then the direction of buckling will be very sensitive to any lateral force applied to the middle of the card.

If you push gently to the left on the middle of the card and then squeeze it, it will buckle to the left. If you push gently to the right and then squeeze, it will buckle to the right. Once the card has begun to buckle in one direction it is difficult to force it in the other direction. In short, a small lateral force applied to the card as an "input signal" can control the direction of the large lateral displacement produced as an "output signal" when the card is buckled by a force applied from the top as a "clock signal".

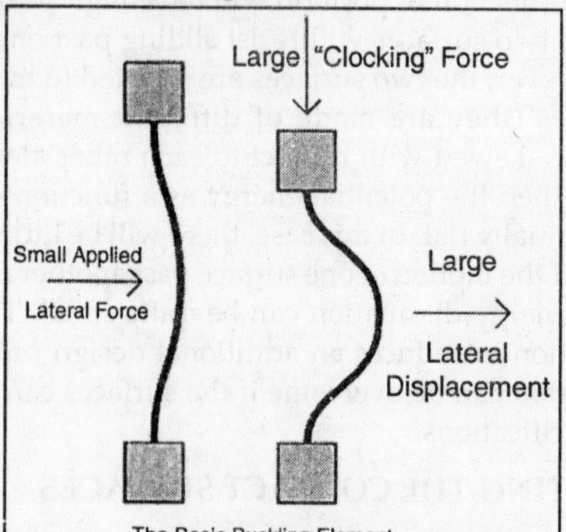

The Basic Buckling Element

This buckling arrangement can be used to define logic operations. If one or more input forces are applied laterally to the midsection of the card, then the midsection will be pushed either somewhat to the left or somewhat to the right, depending on the sum of the applied forces. When the large clocking force is applied, the initial direction of deformation will be continued, and with greater force. This can be viewed as a threshold device, for the output state is controlled by the sum of the input forces.

Once buckled in one direction or the other, the card will remain in that state and will be resistant to a change to the other stable state (in which it is buckled in the other direction). Thus, the device can also be viewed as a "flip-flop" or memory element, as well as a threshold device. This provides a simple mechanical example of a device which, under clock control, can be shifted from a monostable state to a bistable state and back again (e.g., buckled and unbuckled).

Likharev's "parametric quantron" provides an electronic method of doing the same thing. The lateral force applied to the card is analogous to the external magnetic flux phi_e in the parametric quantron, the buckling force (or clock force) is analogous to the control current I_c which controls the critical current IM of the Josephson junction, and the lateral position of the card is analogous to the net magnetic flux phi in the ring.

Likharev said "In fact, the only role of the parametric quantron in our discussion has been to demonstrate how a flexible bistable potential well could be physically realized".

The mechanical implementation given here is another way of making such a flexible bistable potential well which provides a simpler and more accessible illustration of the basic principles. Landauer has previously proposed the use of a bistable spring as an element in a transmission line, and suggested that such a bistable spring could be implemented using a buckled metal plate or card. This proposal was made in the context of signal transmission, rather than computation, and did not consider the need to modulate the spring (e.g., buckle and unbuckle it) for the purpose of computation.

The more general discussion by Landauer of reversible computation using "time modulated potential wells" suggests that the wells could be implemented by the motions of charged particles. The use of buckling cards to implement flexible bistable potential wells and thereby create a reversible mechanical logic system which eliminates surface contact between the parts of the system has not previously been considered. The use of bistable mechanical elements for memory has been proposed by Halg. "It essentially consists of a surface-micromachined bridge which is longitudinally stressed so that it buckles and becomes mechanically bistable.

This bistable bridge performs the memory functions." Their proposal did not go beyond memory, nor consider the concept of buckling and unbuckling the mechanical element to provide a flexible bistable potential well.

The "read" and "write" operations involved a dissipative "snapping" of the buckled element from one stable state to the other stable state. Notationally, the term "well" will be used to refer generically to any flexible potential well which can be shifted from the monostable to the bistable state and back again in a controlled fashion.

In the context of the current paper, "well" usually refers to a mechanical well. Whether we are discussing a generic well or a mechanical well should be clear from context. We will use the phrase "squeeze a well" or "clock a well" to indicate that the well is shifted from its monostable state to its bistable state. The shift from the bistable to the monostable state will be called "releasing" or "unclocking."

Some methods for using wells to build reversible logic have been discussed by Landauer, who described the use of a three-input one-output "voter" circuit. A mechanical implementation of such a device is shown in figure. The three "input" wells are coupled to the "output" well by springs. The springs connect the midsections of the cards in the wells. The springs are bidirectional, and so the distinction between an "input" well and an "output" well is based on which wells have been clocked and which wells have yet to be clocked.

Wells that have already been clocked are "inputs," while

wells that are about to be clocked are "outputs." The three input wells have already been clocked, and the output "voter" well is about to be clocked. Given that all three springs have the same spring constant and have been moved the same distance by the input wells, but that the direction of motion is either to the right or to the left depending on the state of the input wells, then the actual force applied to the midsection of the card of the output well can be approximated by one of the following simple sums: –3, –1, +1, or +3.

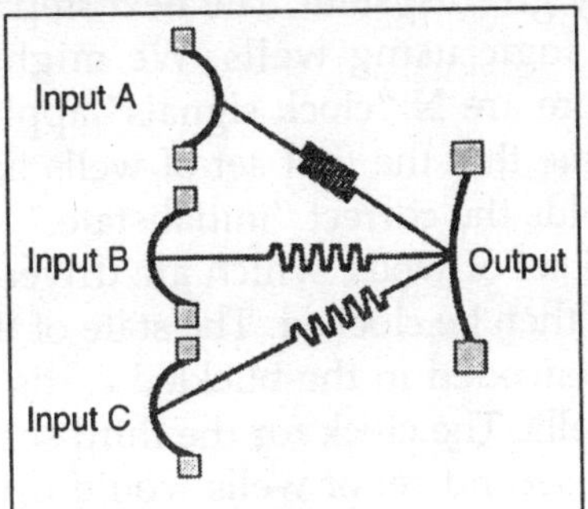

Majority Voter

The card will move to the left if either 2 or 3 of the input cards have buckled to the left, and will move to the right if either 2 or 3 of the input cards have buckled to the right. The input wells are said to "drive" the output wells either left or right through the springs, while the output well is driven either left or right by the springs. "AND" and "OR" gates can be implemented from such voters by leaving one input fixed at logical 0 or 1, respectively.

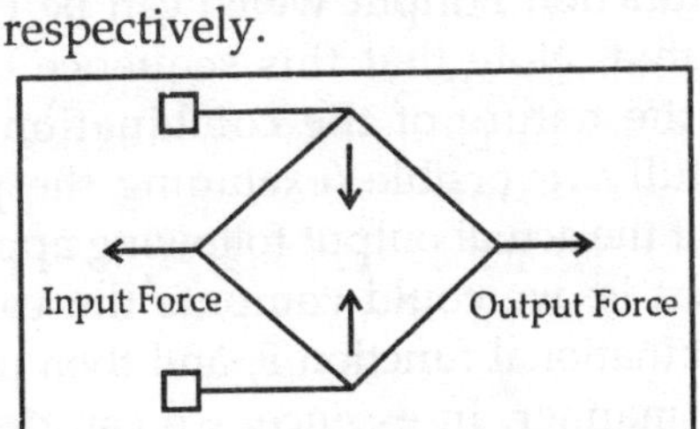

An Inverter

Landauer presumed the absence of a negation operation and so proposed the use of "two rail" logic in which a variable and its complement are available at all times. Negation is then accomplished by exchanging the two variables.

It is relatively simple, however, to make a mechanical inverter and so the current mechanical proposal can be more conventional in this respect. The ability to use an inverter can be thought of as allowing "springs" with a spring constant of –1. More generally, springs with relatively arbitrary strengths can be used. Presumably, the details of the particular implementation will limit the range of values for the spring constants in any given system. The next step is to implement combinational logic using wells. We might, for example, assume that there are N "clock signals" applied to N sets of wells. We assume that the first set of wells has already been clocked, and holds the correct "initial state."

The second set of wells, which are driven by the first set of wells, would then be clocked. The state of the computation would then be encoded in the buckled cards in the first and second set of wells. The clock for the third set of wells would be applied. The second set of wells would drive the third set of wells, and determine the direction in which the cards in the third set of wells would be driven.

This would continue until the Nth clock signal was applied to the Nth set of wells. At this point, the output of the Nth set of wells would be available for use. Once the outputs from the circuit had been used (possibly in a dissipative non-reversible fashion) the clock signals could be removed in the reverse order in which they had been applied. In this way, any arbitrary function F(input wells) can be computed and then uncomputed. Note that this sequence of operations, regardless of the nature of the combinational function F computed, is fully reversible (excluding the possibly non-reversible use of the actual output following application of the Nth clock). That is, we could compute the value of a non-reversible combinational function F, and then un- compute it in a reversible manner. In essence, we can do this because intermediate values in the computation are stored in the intermediate logic levels.

In some sense, the actual "output" is really the state of all the wells in the system, not simply the state of the output wells. Although we're only interested in the state of the output wells and don't care about the state of the other wells, they are available. In the process of reversing the computation, we use the information stored in the wells in all the logic levels, not just the information stored in the output wells.

Put another way, we have embedded the (possibly) irreversible combinational function F(x) into a fully reversible function F′, where the output of F′ includes F(x), x, and all intermediate values used to compute F(x) from x. We will require N "clock signals" to compress the corresponding wells in the proper phase relationship. The simplest way to provide periodic compressive motion is to use oscillators each made from a mass and a spring.

The spring of each oscillator would be connected to a clock driver. The clock driver would simply be a rigid body connected to all the wells associated with the oscillator. The mass would move up and down, compressing and uncompressing the spring. The spring, when compressed, would push the driver down upon the wells, clocking and unclocking them. Unfortunately, such a simple oscillator suffers from two defects. First, the range of motion through which the clock driver moves might exceed what we desire. Second, the duty cycle might not be what we desire.

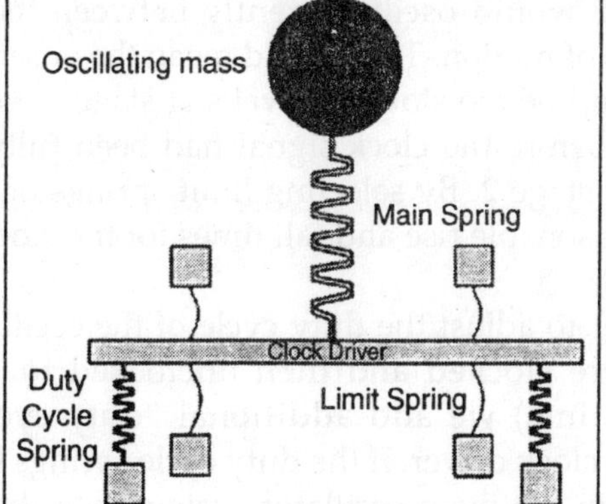

A Clock

In order to achieve good control over the onset and

removal of the clocking force, we add some additional components. In particular, we add "limit springs" both above and below the clock driver. These springs are highly non-linear. When they are pulled and achieve a large extension they flatten out, sharply increasing the force that they apply. Completely straightening the limit springs would require an extremely large force.

This non- linearity can be used to limit the motion of the clock driver to the desired range, e.g., the clock driver will neither crush the cards nor rip off their tops during its normal cycle of operations. Instead, any excess force is taken up by the limit springs. The limit springs have a significantly smaller spring constant than the other springs in the mechanism, and so in most cases either the upper or lower limit springs will be fully extended (e.g., the wells will normally be either "clocked" or "unclocked," and will not occupy some intermediate state).

During the onset and removal of the clock signal, the strength of the limit springs will "soften" the rising and falling edges. If the limit springs had no strength (e.g., were replaced by fixed-length pieces of string) then the clock driver would tend to "snap" from one extreme of its range of motion to the other. This could result in dissipation of energy.

If the spring constant of the limit springs in their linear (non-extended) range of operation were too large, then the clock driver would oscillate gently between its upper and lower limits of motion. This would mean the onset of the clock signals would be too slow: the wells at stage 3 would start to be clocked before the clock signal had been fully applied to the wells at stage 2. By selecting limit springs of appropriate strength, reasonable rise and fall times for the clock signal can be achieved.

In order to adjust the duty cycle of the oscillator (so that the wells are clocked and then unclocked for the correct periods of time) we add additional "duty cycle springs" beneath the clock driver. If the duty cycle springs have greater strength than the main oscillator spring, then the driver will spend most of its time in the up or "unclocked" position.

If the duty cycle springs are weaker than the main oscillator spring, then the driver will spend more of its time in the down or "clocked" position. In the limit, if the duty cycle springs had infinite strength (e.g., were replaced by solid rods) then the driver would always be in the up (unclocked) position. If the duty cycle springs had no strength, the driver would spend half its time in the up (clocked) position and half its time in the down (unclocked) position.

By selecting duty cycle springs with an appropriate strength, the percentage of time spent in the clocked versus unclocked position can be controlled.

These clocking mechanisms involve nothing but simple mechanical motion with no contact points. By adjusting the duty cycle and the phase relationship of the different oscillators, the proper clocking signals can be generated to drive the wells.

Once started, the oscillators will tend to drift and would have to be kept in the proper phase relationship with each other. Ideally, they would be phase locked. If, for example, N clock oscillators were operating at some frequency f and small multiples of f, then we might have a single master oscillator at a frequency that is a small multiple of Nf.

The clock oscillators would all be phase locked to the master oscillator, and so the proper phase relationship would be preserved. Because each clock oscillator can control an arbitrary number of wells, and because the oscillators should drift out of phase rather slowly, the energy (if any) dissipated in maintaining the proper phase relationship can be amortized over many well operations.

The energy dissipation per well operation from this source can be made arbitrarily small. If we were to erase information, we would expect to find an energy dissipation mechanism which dissipated ln(2) kT or more per bit erased. We would not be disappointed.

If, for example, we were to unclock logic level 1 then it would be necessary for the input to that logic level to be identical to the actual information already stored in the logic level if we are to avoid energy dissipation.

In other words, each well in logic level 1 will be driven either left or right by its input. If some particular well in logic level 1 is buckled left but is being driven to the right by its input signal, then unclocking logic level 1 will be a non-reversible operation. Once we release the clock, the well will forget its original state. Should we at this point reapply the first clock, the card would buckle to the right in response to its input and not to the left: we have forgotten the original state and hence performed an irreversible act. We must pay for this forgetfulness with energy.

The energy loss in this particular system occurs as we remove the clock signal. A well buckled to the left but driven to the right will, at some point during removal of the clock, "snap" from left to right. This "snap" will occur when thermal noise is sufficient to allow the card to overcome the shrinking energy barrier that separates the leftward position from the rightward position.

It will indeed dissipate an amount of energy which is several times kT. To prevent such dissipative "snap" losses, we must never drive a buckled well in the wrong direction and unclock it.

If we do, we will suffer energy losses which are several times kT. Even more strongly, if we wish to unclock a buckled well the buckled well must be driven in the *correct* direction during the unbuckling operation. In this discussion, we assume that a well is always driven in one direction or the other. The only time that a well would not be driven in either direction would be during the very brief transition when the direction it was being driven was changing from left to right or from right to left.

Thus, all unbuckling operations will take place with the well driven either left or right. While our objective is to design a reversible logic system in which energy dissipation can be completely avoided, we might find it convenient on occasion to erase a bit. In this case, we know that the least energy dissipation we can hope to achieve is $\ln(2)\, kT$.

We can approach this limit by unclocking the well when the driving force is 0, e.g., the well is driven neither left nor right. In this way there will be no "snap" as the well is unclocked. This zero driving force can be approached by connecting a well that we wish to erase (the "erasee") to an otherwise unused "erasure well." The information held in the erasee can be moved into the erasure well by clocking the erasure well. Then we can "unwrite" the erasee without dissipating energy.

Hence, when we unclock the erasure well, the driving force on it will favour neither the left nor the right position, as desired. Therefore, in those (presumably infrequent) cases where we do decide to use an irreversible operation, it can be made to dissipate close to the minimum amount of energy possible. As mentioned in the previous paragraph, in this discussion we are focusing on reversible logic which does not erase information, and so will not consider unbuckling a well when the driving force is 0.

Given a method of computing combinational logic, we must still build a computer. Again, as with non-sliding logic, we assume that there is a reversible function F whose iteration is useful. We can use exactly the same general method as described before. We add an additional logic stage to hold the output of the Nth logic level. Again, we refer to the Nth logic level as the "output," and the N+1st level as the "latch" stage. In this case, the latch stage will be identical with all preceding stages. It is distinguished only in that each "latch" has a single input from the output stage, rather than the three inputs that are common in the earlier logic levels.

In this way, the wells in the "latch" stage will be buckled in exactly the same direction as the wells in the output stage. The springs that carry force between the output stage and the latch stage will always drive the output wells in the correct direction (thus eliminating the possibility of a "snap" mechanism of energy loss). We can then safely unclock the output logic level and all preceding logic levels back to the 1st logic level - which must remain clocked until it can be "unwritten." The rest of the construction used with non-sliding logic is now used for buckling logic, and results in a device which is asymptotically nondissipative.

In general, the states of an individual well can be described as follows: the well is either clocked and buckled left, clocked and buckled right, or unclocked (and unbuckled). Further, the driving force applied to the card of the well can be either left or right (we ban the application of 0 force, as achieving exactly 0 force might be difficult in practice. This is certainly a pragmatic issue in many systems, but it is unclear if it is fundamental). The well can therefore be in one of six states. If buckled left, buckled right, and unbuckled are abbreviated “(“, “)”, and “|”; and if driven right and driven left are abbreviated “>” and “<“, then the six states are: “>(“, “>|”, “>)”, “(<“, “|<“, and “)<“.

The symbols “>(<“, “>|<“, and “>)<“ will represent the three sets of two states in which the drive force is either left or right, e.g., “>(<“ is the set of two states {“>(“, “(<“}. The allowed transitions are: (< to |<; |< to (<; >) to >|; >| to >); >(< to >(<; >)< to >)<; and >|< to >|<. Any transition which is not allowed is forbidden. Among the banned transitions: a card that is buckled left cannot suddenly shift to buckled right, so >(< cannot become >)<, and >)< cannot become >(<.

This is shown in the following table:

<table>
<tr><th></th><th></th><th colspan="6">Final State</th></tr>
<tr><td rowspan="6">Initial State</td><td>>(</td><td>+</td><td>+</td><td>–</td><td>–</td><td>–</td><td>–</td></tr>
<tr><td>(<</td><td>+</td><td>+</td><td>–</td><td>+</td><td>–</td><td>–</td></tr>
<tr><td>>|</td><td>–</td><td>–</td><td>+</td><td>+</td><td>+</td><td>–</td></tr>
<tr><td>|<</td><td>–</td><td>+</td><td>+</td><td>+</td><td>–</td><td>–</td></tr>
<tr><td>>)</td><td>–</td><td>–</td><td>+</td><td>–</td><td>+</td><td>+</td></tr>
<tr><td>)<</td><td>–</td><td>–</td><td>–</td><td>–</td><td>+</td><td>+</td></tr>
</table>

Allowed (+) and forbidden (–) transitions

Also, to prevent undesired energy loss, we cannot unclock a well that is buckled left but driven right, nor one that is buckled right but driven left: we ban the transitions >(to >| and)< to |<. A few other unreasonable transitions are also forbidden.

An unclocked well can actually transmit the drive force through itself. That is, some fraction of the drive force from the preceding well(s) can be transmitted through to the following wells. These wells will, in their turn, transmit the (now much reduced) forces on to other wells.

In general, a complex web of diminishing forces will be transmitted through the unclocked wells in a system. However, if the unclocked wells have a vertical upwards force applied through their drivers, then the lateral force transmission will be substantially damped on passage through each well.

Because the "legitimate" forces are significantly larger and take on only a few nominal values, small variations in the actual force applied to a particular well will not cause it to buckle the wrong way when clocked. The course of the computation is not affected by small variations or perturbations in the system.

It is clear, however, that a significant figure of merit is the extent to which an unclocked well will prevent through-transmission of an applied driving force. The less force that is transmitted, the lower the "noise" forces in the system.

Reduced noise will allow larger fan- in and fan-out. (Interestingly, these "noise" forces can be usefully employed in a different logic mode, described later, and the "figure of merit" would become a "figure of demerit.")

Once the clock oscillators are adjusted so that the oscillations have the correct amplitude and phase relationships, the computation will proceed. By reducing the speed of operation, the energy dissipation per well operation can be made arbitrarily small.

A perfectly elastic material would exhibit no losses. Real materials approach this ideal as they move more slowly, so energy dissipation from this source can be reduced indefinitely. The method of maintaining the proper phase relationship between the oscillators might possibly require energy, but the energy dissipation per well operation (if any) from this source can be made arbitrarily small by increasing the number of wells per clock phase (only a small, finite number of clock phases are required even when the number of wells is increased indefinitely); and by increasing the stability of the oscillators so that adjustments in their frequency are infrequent.

Other sources of energy dissipation (e.g., electromagnetic radiation) appear to be insignificant by comparison. It is instructive to consider the system from the point of view of a single atom within it. The local neighborhood around the atom would change very little during each cycle: only small perturbations in the local geometry would take place.

While large macroscopic motions would occur, these would never involve large changes in local structures, e.g., they would never involve the close approach of two atoms that were initially far apart, nor the separation of two atoms that were initially close. The local potential energy function describing the energy of the local arrangement of atoms as a function of the positions of their nuclei would be in a local minima. The small perturbations in position caused by the elastic deformations of the overall structure would dissipate very little energy if they were not too rapid.

The energy loss that must inherently take place in such a system can be made arbitrarily small by reducing the speed of operation. A possible candidate for a very small card would be a sheet that was one or two atoms thick. A sheet that was 10 nanometers high by 1 nanometer wide would have a few hundred atoms. We will use a very much larger size (10,000 atoms) in the calculations below simply to provide a large design margin. That is, it seems virtually certain that some arrangement of 10,000 or fewer atoms will correctly implement a card and related connecting springs.

If we reduced the number of atoms available to build the device, then we would need to specify the device design in greater detail to insure that the construction was still feasible. (As an example of what should be possible in purely mechanical systems, a single "lock" in Drexler's rod logic has a volume of 4.1 cubic nanometers and somewhat fewer than 500 atoms).

From a mechanical point of view, a card will become increasingly sensitive to lateral force applied to its midsection as it becomes thinner and longer. Quantitatively, we have

$$d = f h^3 / (4 E t^3 w)$$

where f is the lateral force applied to the middle of the card, h is the height of the card, E is Young's Modulus, t is the thickness of the card, w is width of the card, and d is the distance the centre of the card moves under the applied force. The continuum approximation will break down as the card becomes sufficiently small and atomic structure becomes significant.

In this case, molecular modeling techniques that take into account bond bending and other bond parameters would be required to analyse the behaviour of the structure.

In point of fact, a card under compression is a tristable device. That is, it can buckle either to the left, to the right, or balance (perhaps precariously, depending on its thickness and height) straight up and down. As the card becomes thinner and higher, the size of the "balancing" region will become smaller and smaller. In the case of a card that was one atomic monolayer thick, this region would be vanishingly small. In a logical sense, this has no impact on the previous discussion.

If we have a tristable device, then we can simply exclude one of the three stable states and utilize the remaining two stable states for device function. While this makes the "feel" of the terminology we have adopted somewhat odd ("buckled left," for example, might be implemented by a card that is balanced under compression) the principles still apply.

Alternatively, by selecting the lateral force large enough to drive the physical card into either the left-most or right-most stable state (avoiding the central stable state entirely) we can (for logical purposes) ignore the existence of the central stable state. Of course, this will reduce the force sensitivity of the device (unless the card is an atomic monolayer, in which case the central stable state does not exist). On the other hand, we could get maximal lateral force sensitivity by using the central stable state and the right stable state (using only two of the three physically available stable states). A device that used only two of the three physically available states should exhibit remarkable sensitivity to lateral force. If it were properly balanced (either by being manufactured that way to begin with, or by being "trimmed" afterwards) the only limiting factor on the force sensitivity would be thermal noise.

That is, the energy spent in pushing the card laterally would have to exceed several times kT, for thermal noise might otherwise cause the card's position to fluctuate enough to cause it to buckle in the wrong direction. There seems to be no other inherent lower limit on the applied lateral force.

Even though buckled logic eliminates several obvious sources of energy dissipation, dissipative mechanisms still exist. The mere fact that a real elastic system is buckled and unbuckled will cause energy dissipation in much the same way that compression and decompression of a gas causes energy dissipation.

Drexler concluded that the major source of energy dissipation in rod-logic operating at 50 picoseconds was from this kind of mechanism. The fundamental energy dissipation mechanisms that exist even in a perfect non-conducting solid are thermoelastic damping and phonon damping. An infinitely slow isothermal cycle of compression and decompression of a gas can be dissipation free. In the same way, the isothermal buckling and unbuckling of a card can also be dissipation free. In reality, of course, neither operation is infinitely slow nor perfectly isothermal. If a single cycle takes time t, then the energy dissipated per cycle will be delta E = k/t for some suitable constant k. By slowing down the cycle time, the energy dissipation per cycle can be reduced indefinitely.

It is interesting to note that a relatively rigid structure of small size can only support vibrations above a certain frequency. If the speed of sound in the structure is 10,000 meters per second while the size is 10 nanometers, then the lowest frequency vibrations will be about 10^{12} hertz. Quantum theory dictates that the allowed vibrational energies of a vibrational mode at frequency f are $(n+1/2)\, h f$, where n is the quantum number and h is Planck's constant or 6.6×10^{-34} joule seconds. Thus, the vibrational energies are separated by $h f$ or about 6.6×10^{-22} joules. This corresponds to a temperature of $6.6 \times 10^{-22}/1.38 \times 10^{-23}$ or almost 50 kelvins. The average value of n at thermal equilibrium, <n>, is given by:

$$\langle n \rangle = 1/(e^{h\,t\,/kT} - 1)$$

At 50 kelvins $\langle n \rangle$ is about 1.7. At 2 kelvins, $\langle n \rangle$ is about 10^{-11}, so the probability that n equals 0 is almost 1 (e.g., the system is almost certainly in the ground state).

To rephrase this in words, at some sufficiently low temperature a small rigid structure has essentially zero heat capacity because the transition from the ground state to the lowest excited state will require significantly more energy than can be provided by thermal noise. This in turn implies that buckling and unbuckling (e.g., mechanical distortions of the structure) will not cause any heat flow. This raises the interesting possibility that operation of a mechanical logic device below a certain temperature can result in an exponential decrease in energy dissipation as a function of decreasing temperature.

A logic device that dissipated such a small amount of energy is not in principle impossible, so this possibility (while remarkable) does not violate any physical laws.

An actual mechanical computer would have many gates, so the size of the whole structure could be quite large. As a consequence, lower frequency vibrational modes would exist in the overall structure. If the wavelengths associated with these lower frequencies are significantly larger than the size of an individual logic element, then it should be feasible to make coupling between logic operations and these low frequency modes very small. Further analysis of the possible energy dissipation mechanisms involving these lower frequency modes is needed before firm conclusions can be made.

Measurements of energy dissipation of single-crystal silicon (including measurements at temperatures of 0.005 to 4.2 Kelvin) by Kleiman et. al. show that Q's in the range of 10^4 to 10^8 can be achieved (e.g., an energy dissipation of one part in 10^4 to one part in 10^8 per cycle).

They identified two-level systems, possibly created by oxygen and carbon impurities in the sample, as the dominant energy dissipation mechanism at low temperature. Their samples were also relatively large and as a consequence they were operating in a regime where *h* omega << *kT* (e.g., low frequency phonons with energies much lower than thermal noise were present), so their results are not applicable to questions of energy dissipation by a small device (in which low frequency phonons do not occur).

If the imperfections of the material were eliminated and the size reduced, substantial reductions in energy dissipation should be feasible. Buser and Derooij achieved a Q of over 600,000 for Si at about 24,000 Hz. Hu et. al. obtained a Q of 400,000 for SiC at 4 K at a frequency of about 20,000 Hz . Quartz crystals with a Q of greater than 10^9 (a billion) at a temperature of about 2 Kelvins and a frequency of a megahertz have been observed experimentally.

Using bulk approximations, in theory 2 PI fQ = 10^{14} for quartz oscillators at room temperature if defects in the material are eliminated. At one megahertz, this implies a Q of greater than 10^7.

Energy dissipation can also be caused by excitation of vibratory motions by the clocking motion itself. In this case, the energy dissipation will depend on the frequency spectrum of the applied clocking motion.

If some vibratory mode has a high frequency, it will not be excited by a clock signal which has no high frequency components. By this logic, the optimal profile for the clocking motion is a pulse which has as few high frequency components as possible. The best candidate is the gaussian function exp(–px^2). When x = 0, the gaussian has a value of 1.

As x increases or decreases, the value of the gaussian falls off smoothly and rapidly to very low values. The fourier transform of the gaussian is also a gaussian: exp(-ps^2), where s is the frequency.

The gaussian has a faster than exponential fall off in amplitude at higher frequencies. If we start with simple ideal square-wave clock motions, they can be convolved with a gaussian to produce a "softened" square wave with gradually rising and falling edges. The resulting clock function will retain the faster than exponential fall-off of high frequency components.

By employing a gaussian-softened time profile for the clocking motion, the energy dissipation of a logic element can be reduced faster than exponentially by reducing the frequency with which this clock pulse is applied to significantly below the frequency of the lowest vibratory mode of the logic element.

In this way, excitation of any vibratory modes can be avoided. As noted above, a reasonable low- frequency cutoff for a small logic gate of some 10,000 atoms is in the picosecond range. Nanosecond clock pulses which do not contribute any significant amount to energy dissipation by this mechanism should therefore be possible.

As a consequence of these considerations we conjecture that there exists a physically realizable system (e.g., an appropriately arranged collection of atoms) which implements the general purpose capabilities of a computer and in which the energy dissipation per logic operation falls off exponentially as both the temperature and the speed of operation are simultaneously reduced.

This conjecture is qualified by the observations that, at some point, energy dissipation from gravity waves (incredibly slight as this would be in such a system) would eventually become significant if all other sources of energy dissipation were sufficiently reduced; and by the further observation that radiation caused by the acceleration of charged particles, even though the mechanical systems considered here could be implemented by very nearly charge-neutral structures with extremely small dipole moments, would also eventually become significant.

Further research on the energy dissipation modes involved in systems of many logic elements is required to confirm or refute this conjecture. "Buckled logic" seems to be an evolutionary endpoint in efforts to reduce the interactions between the parts of a mechanical logic system.

The only interactions are elastic. It provides a simple mechanical model for a class of clocked reversible computations, and is another illustration of the fact that a clocked mechanical reversible computer which dissipates an arbitrarily small amount of energy per logic operation can be built from an appropriate arrangement of atoms.

RELATIONSHIP OF BUCKLED LOGIC AND ROD LOGIC

The relationship between buckled logic and rod logic might not be immediately obvious. In fact, rod logic is also a member of the broad class of clocked reversible computational models which use a flexible potential well to perform logic operations. The gate knob in rod logic blocks the freedom of motion of the probe knob, and so prevents the probe rod from freely moving through its entire range of motion.

The "potential well" in which the probe rod moves is thus bifurcated into two separate regions by motion of the gate knob. There are three main conceptual differences in rod logic. First, the potential well has a flatter bottom and steeper sides, changing the intuitive "feel" of its behaviour.

Second, the "coupling constant" between wells is much higher. Finally, because of these two differences, Drexler employed a logically different method to map abstract logic functions onto a physical device.

As an example, we can consider how to use potential wells similar to those used in buckled logic to implement logic functions in the rod logic style. Consider two wells, A and B. If we set the spring constant of the spring that couples A and B to infinity (e.g., we replace the spring with a stiff rod) then the cards in the two wells will necessarily have the same lateral motions. If we apply a lateral force to the card of well A, this necessarily is applying a lateral force to the card of well B, for they are now one mechanical unit.

For this reason, when we apply a lateral force we can simply say that it is applied to the rod and need not attempt to distinguish which card of which well it has been applied to, for it has been applied to both. Should we apply a leftward lateral force to the rod, and then buckle either A or B, the rod will be locked into a leftward position. A subsequent rightward force will not change the position of the rod appreciably. If, however, we buckle neither A nor B, then a subsequent rightward force will move the rod to the right. In essence, this is an "AND" gate.

Instead of treating the buckling force as a clock input, we are treating it as a signal input. Instead of treating the lateral force as a signal input, we are treating it as a clock input. Thus, the roles of "clock" and "signal" have been reversed. If we replace "squeeze the well" with "apply force to move the gate knob into the lock," and if we replace "apply a rightward force to the rod connecting the wells" with "pull on the rod" then the essential relationship of buckled logic to rod logic can be seen.

The possible combined states of two wells connected by a rod are: >((<, >| |<, >))<, >|)<, >)|<, >|(< and >(|<. The states >()< and >)(< can reasonably be banned. State transitions can be defined as before. Just as there can be many gates on a rod, so there can be many wells connected by a rod. Three, four, or more wells can be connected by a rod, representing multi-input AND gates. Different physical devices, which have differently shaped potential wells and different coupling characteristics between wells, can use either approach. Some devices will lend themselves better to the rod logic approach, others will be better served by the buckled logic approach. Some devices can be adapted for use in either fashion.

REVERSIBLE ELECTTHE CONTINUING

Revolution in computer hardware is packing ever more logic gates in ever smaller volumes. Unfortunately this produces ever more heat, which will limit the feasible packing density and performance unless the energy dissipated by each logic operation can also be dramatically reduced. To reduce the energy dissipation of a logic operation below ln(2) kT (near thermal noise) requires the use of reversible logic for fundamental thermodynamic reasons. Extrapolation of current trends suggest this limit will become significant within one to two decades.

Many real devices can be viewed as electrically controlled switches, so a method of using an abstract switch in a reversible manner is useful. Two methods of using switches to implement reversible computations are discussed in this paper. The first method has an energy dissipation which is proportional to the square of the error in the voltage, while the second method has an energy dissipation which can in principle be reduced indefinitely by slowing the speed of computation.

The first method is basically an extension to "pass logic" which has been previously used with both nMOS (hot clock nMOS) and CMOS transmission gates to achieve low energy dissipation. The second method is a novel thermodynamically reversible logic system based on CCD-like operations which switches charge packets in a reversible fashion to achieve low energy dissipation. There is now a fairly extensive literature on reversible computation which shows that the energy dissipation per device operation cannot be reduced below ln(2) kT (where k is Boltzman's constant and T is the temperature; kT is roughly the thermal energy of a single atom) if the device is not reversible.

For the last 50 years the energy dissipation per gate operation has been declining with remarkable regularity. Extrapolation of this trend shows the energy dissipation per device operation reaching kT by the year 2015.

This assumes that T is 300 Kelvin - more on this later. To gain some perspective on this consider that an "AND" gate which has a power supply of one volt and which allows 100 electrons to go from that one volt supply to ground during the course of a switching operation will dissipate 100 electron volts. Although 100 electron volts is about 4,000 times kT (and well above the theoretical limit), it will be difficult for simple improvements of irreversible approaches to reach even this level of energy dissipation.

Extrapolating present trends, we should reach 4,000 kT before the year 2000, e.g., within ten years. While reversible logic is an absolute necessity if we are to reduce energy dissipation per device operation below kT, it is a useful heuristic for designing systems that have low energy dissipation even when the actual energy dissipation is well above kT.

Even if we do develop irreversible devices that approach ln(2) kT, a computer operating at room temperature at a frequency of one gigahertz with 10^18 logic gates packed into a volume not much larger than a cubic centimeter would dissipate over three megawatts. The drive for ever greater computational power with ever more densely packed logic elements will eventually require that a single logic operation dissipate orders of magnitude energy less than kT.

If we are to realise the full potential of nanoelectronics and molecular logic devices, at some point in the future we will be forced to use thermodynamically reversible logic elements for a substantial fraction of logic operations.

We can state quite confidently that one of three things will occur:

- The historic rate of decreasing energy dissipation per device operation will slow or halt in the next one or two decades
- We will operate computers at lower temperatures or
- We will develop new methods of computing in a reversible way that can beat the kT barrier.

The first option is unattractive. First, cooling is a major problem. The heroic cooling methods used in the Cray 3 supercomputer to remove the heat generated by the computer's operation suggest that failure to reduce energy dissipation per gate operation would be a major limiting factor in future computer performance.

While both air-cooling a single chip that dissipates 150 watts and water cooling of a chip dissipating 790 watts/cm^2 have been demonstrated , and more effective cooling methods should be feasible in the future, we will eventually reach a limit to heat removal.

Second, in many applications power is limited. Portable computers, remote sensors and other isolated systems have only a limited amount of power available. Third, although the raw cost of electrical power is not yet a major limitation it would become so in the future if reductions in energy dissipation did not keep pace with advances in other areas.

The Wall Street Journal said "Computer systems currently account for 5% of commercial electricity use in the U.S., with the potential to grow to 10% by the year 2000." Operating computers at a lower temperature will not reduce overall energy dissipation. If we operated future devices at 3 Kelvins we would reduce kT by a factor of 100 and so could reduce energy dissipation per logic operation by a similar factor - but for fundamental thermodynamic reasons the coefficient of performance of the refrigerator for the system can be at best 3K/(300K – 3K) or about 0.01.

Thus, the lower energy required per gate operation will be almost exactly balanced by the increased energy needed by the refrigerator. In many applications refrigeration is not an attractive option. Lap top and portable computers, embedded systems, various "smart" appliances and other applications make the use of refrigeration undesirable.

Factors other than net energy savings can make low temperature operation worthwhile. For example, some potentially attractive devices don't operate at higher temperatures.

In large computers operating in stable environments (the traditional computer centre, for example) refrigeration might be attractive, particularly if it permits the use of devices that provide much better performance but which require low temperature for their operation (Josephson junction devices might be an example).

Refrigeration per se, however, does not seem too attractive. Finally, and most attractively, we could use reversible computation to reduce energy dissipation. This would, in principle, allow energy dissipations indefinitely below kT per logic operation.

While some barrier should eventually be encountered, the use of reversible logic should allow us to continue current trends in energy dissipation per logic operation for the longest possible time. It is interesting to note that certain uses of pass logic, transmission gates, and hot clock nMOS do in fact perform some logic operations in a thermodynamically reversible fashion.

Research in the design and utilization of reversible logic to futher lower energy dissipation will simply recognize and make explicit a trend that has already begun. Further research in this area is the most appropriate response to the rather limited range of possibilities that face us.

REVERSIBLE ARCHITECTURES

One concern about this approach is the need to use reversible computer architectures. Such architectures are entirely feasible. A wide variety of computations can be done in a logically reversible manner. Bennett concluded: "...for any $e > 0$, ordinary multitape Turing machines using time T and space S can be simulated by reversible ones using time $O(T^{(1+e)})$ and space O(S log T) or in linear time and space $O(S\ T^e)$." This general result shows that even arbitrary irreversible computations can be mapped into reversible computations.

Specific irreversible computations can often be mapped into reversible computations with little or no loss of efficiency. Even if we do not adopt new reversible computer architectures, simple applications of reversible computation could be made within the framework of existing architectures.

A typical computer executes a sequence of instructions, and each instruction will typically change the contents of a single register or memory location. Although loading the result of the instruction into a register will normally be irreversible (it erases the previous contents of the register) it is still the case that all other operations performed by the computer during instruction execution could in principle be made reversible.

Thus, although we would have to dissipate roughly kT energy for each bit in the output register for each instruction execution, we need not use dissipative logic throughout the computer (as is currently done).

We can go one step further without making significant changes in computer architecture: the simple register-register add instruction $R_1 = R_1 + R_2$ is logically reversible. With proper hardware, this particular instruction could in principle be made to dissipate as little energy as we desired.

While non-reversible instructions (e.g., $R_1 = 0$) would still dissipate greater energy, the overall energy dissipation of the computer could be reduced significantly if a reversible approach were developed and used to implement the reversible instructions.

Of course, once the energy-wasting instructions were identified, compilers would learn to avoid using them. Today, compilers can optimize for speed of computation and for the amount of memory used. In the future, compilers could also optimize a programme to minimize the energy dissipated during computation, e.g., to minimize the number of bits that are erased. This would provide an entirely evolutionary path from the current irreversible designs to computer architectures that were as reversible as was practically feasible.

While it is perhaps not obvious how far this trend can go it is clear that a very large percentage of computer operations can be made reversible - perhaps a remarkably large percentage.

REVERSIBLE DEVICES

Several reversible devices have been proposed over the years. Von Neumann and independently Goto proposed the parametron which encodes information in the phase of an oscillation.

Fredkin and Toffoli proposed an electronic reversible logic family which used switches, capacitors and inductors. By turning the switches on and off at just the right times, charge could be transferred between the capacitors through the inductors with minimal energy loss.

By increasing the size of the inductors it would be possible to create a family of ever slower but ever more energy efficient circuits. As the LC time constant is increased, the energy dissipation can be decreased to an arbitrary extent.

A proposal by Likharev based on Josephson junctions "...is particularly significant, because it is a genuine example... of a system that has frictional forces proportional to velocity" according to Landauer.

The parametric quantron is based on Josephson junctions and operates at low temperatures. It should dissipate less than kT when operating at 4 kelvins and with a switching time of 1 nanosecond. Modern "high temperature" superconductors might allow operation at relatively high temperatures, including that of liquid nitrogen.

Even these temperatures would still be a significant disadvantage. Molecular-scale mechanical reversible "rod logic" has been proposed by Drexler which should dissipate less than kT when operating at room temperature at a speed of 50 to 100 nanoseconds. Drexler's proposal is also significantly smaller than the parametric quantron. While the ability to fabricate Drexler's proposal is some decades away, it provides a good argument that molecular-scale reversible devices should be feasible.

Although the parametric quantron can be used in a reversible fashion to achieve low energy dissipation it uses magnetic fields to couple logic devices, requires low temperatures to operate and will likely prove resistant to scaling to the molecular size range.

By contrast, Drexler's mechanical proposal might well prove smaller than even the smallest of future electronic devices.

Mechanical devices depend fundamentally on the position of the nuclei of the atoms of which they are made. The position of the nucleus of an atom can be much more precisely known than the position of an electron: quantum uncertainty in the position of the electron is much greater because the mass of the electron is much smaller.

If device function depends on the position of an electron then device size will be larger than if device function depends on the position of the nucleus. Such molecular mechanical proposals will, however, almost certainly be slower because of the greater mass of the nuclei of atoms as compared with the mass of electrons.

In this paper we discuss simple methods of using relatively conventional electronic switching devices in a reversible fashion which can be scaled to a very small size. The methods described here allow a simple voltage-controlled switch to be used in a reversible manner, both to implement combinational logic and also to iterate an arbitrary reversible logic function. Other methods are feasible. For example, helical logic — an elegant form of reversible electronic logic — has recently been proposed by Merkle and Drexler.

Voltage controlled switches of various types have long been used in electronic devices. Not all voltage controlled switches are suitable for use in low energy designs because of their intrinsic energy dissipation. Traditional relays actuated by electromagnets dissipate comparatively large amounts of energy.

Even if a switch were suited for low energy operation it would still dissipate energy if it were turned on when there was a voltage across it. The abrupt rush of current that normally occurs when a switch is turned on would insure this. Most previous logic designs based on switches operated in this manner (though the proposal by Fredkin and Toffoli avoids this problem, and the proposal by Seitz et. al. eliminates dissipative switch closures for combinational logic, although not for sequential logic).

The switch is turned on for the purpose of draining charge when the charge is no longer wanted. That is, a logic "1" might be encoded by a charge on a capacitor. Turning on a switch to drain away that charge would dissipate energy even if the switch itself were perfectly non-dissipative.

A related but distinct line of research also emphasizes the importance of energy dissipation, but assumes that the losses involved in dissipatively switching a wire from one logic state to another are fixed and cannot be altered, e.g., that slowing the switching speed will not reduce energy dissipation.

Even assuming that this is true, clever design of the circuit can greatly reduce the number of times it is necessary to take such a dissipative action. Further research which combines both this approach and the approaches described here would seem fruitful.

Even relatively recently some authors have argued that switching devices must fundamentally dissipate more than kT regardless of how slowly they operate. It is worth noting, therefore, that charging and discharging a condenser need not dissipate the energy involved. For example, a simple LC oscillator will repeatedly charge and discharge a condenser.

Although the energy stored on the condenser is 1/2 CV^2, this energy need not be dissipated during a charge/discharge cycle.

If the Q of the circuit is high the energy dissipation per cycle will be small. In addition, the energy dissipation of such an oscillator depends on frequency. By using larger condensers and inductors the frequency of operation can be slowed and the fraction of energy dissipated per cycle reduced for a discussion of scaling laws). If all linear dimensions of a device are scaled in size in proportion to some characteristic length L then capacitance and inductance are proportional to L while frequency and resistance are proportional to 1/L.

The Q of the circuit is proportional to L so the percentage of the total energy in the oscillator that is dissipated per cycle can be reduced simply by increasing all linear dimensions equally. Different scaling methods can produce better results. If we assume that a fixed capacitive load is charged through a fixed resistance by a voltage that is ramped up over a period of time T, then the energy dissipated by the resistance is approximately 1/2 $CV^2 \times 2RC/T$ (for $RC \ll T$).

This dissipation can be reduced as much as we might want by increasing T (and leaving the circuit components unchanged). Koller and Athas have called this adiabatic charging. We will call it "thermodynamically reversible" or simply "reversible," relying on context to make it clear that we are referring to an asymptotically non-dissipative process. Note that the term "reversible" can also be used to mean "logically reversible" with no implication that it is asymptotically non-dissipative.

A logically reversible process might be highly dissipative, so it is sometimes necessary to distinguish between "logically reversible" and "thermodynamically reversible." The term "reversible" in this paper will usually mean "thermodynamically reversible and asymptotically non-dissipative." The reversible charging of a condenser is illustrated in figure. The situation is quite different for a step function in which the voltage is ramped up instantly (T=0). This process dissipates 1/2 CV^2. Charge falls through the fixed potential drop in a dissipative fashion much as water goes over Niagra falls. This is illustrated in figure . Although this is the method in common use today to charge and discharge the various capacitive loads in computer circuits, it is not the method that we will be considering.

Comparing figures 3 and 4 shows that the advantage of reversible charging is in the factor of 2RC/T. Normally the switching time T must be longer than the RC time constant of the circuit simply to insure that the output has time to settle to its proper value.

By using reversible charging, further increases in the switching time can be used to decrease the energy dissipation per logic operation. This, of course, results in a slower speed of operation leading to a trade-off between speed and energy dissipation. How to arrange matters so that current flows smoothly and reversibly is the principle topic of the rest of the paper.

This is challenging because the basic switching devices are:

- Abrupt and discontinuous
- Pass through a dissipative intermediate state (which is neither fully on nor fully off) while switching.

THE FIRST METHOD

The first method uses a conventional voltage controlled switch but in a somewhat unusual manner. In essence we never turn a switch on or off when there is either a voltage across it or current going through it. This mode of operation is reversible.

This approach has previously been used to compute combinational functions e.g., appropriately implemented pass logic or CMOS transmission gates. We extend this approach to implement a sequential circuit which can iterate the computation of a reversible logic function. Several authors have pursued this approach recently, including van de Snepscheut, Koller and Athas and Hall and have reached similar conclusions. It is also related to "dry switching," a method for using relays which avoids turning relays on or off when there is a voltage across them. The primary purpose of dry switching is to avoid deterioration of the relay contacts, rather than minimize switching energy. Conceptually, a voltage controlled switch can be thought of as a relay which is controlled electrostatically, as illustrated in Figure. We adopt the symbol shown in Figure for a voltage controlled switch. We also adopt the following convention for turning a switch on and off: when the control input to the switch is at logic "0" (low) the switch is turned on and current can flow from the input to the output or from the output to the input.

When the control input is "1" (high) the switch is turned off and current cannot flow. The reader should note that this convention is not the usual one. We adopt it instead of its opposite —where a "1" turns the switch on and a "0" turns it off— because it is more intuitive when applied to the second proposal discussed later in this paper, the CCD-based reversible logic element. The proposed mode of operation is fundamentally clocked and multi-staged. The outputs of stage i drive the inputs of stage i+1. For the moment, we assume the output of stage 0 is simply given. That is, a set of input lines are assumed to have fixed voltages upon them representing the input data.

The value of the succesive logic levels can then be computed in turn. The logic values (voltages) produced by stage 1 are computed from the logic values given by stage 0; The logic values produced by stage 2 are computed from the logic values produced by stage 1; etc.

Simple pass logic can be employed to implement combinational logic. A "NOR" operation in this logic is simply two switches in series. A clock signal (initially at logic 0) is connected to the input of the first switch.

If the control inputs of both switches have settled to logic 0, then both switches will conduct. The clock input connected to the first switch is slowly raised (from logic 0 to logic 1). Both switches will conduct and the output will gradually shift to a logic 1. If either input is 1, one of the switches will not conduct and the output will remain at 0.

N-LEVEL CIRCUIT

We assume that we have an N level combinational circuit,

and that N "clock signals" are provided. Initially, clocks 1 through N are low (logic 0), and the outputs of stage 0 are at some preset values. The outputs from stage 1 are computed as simple logical functions from the outputs of stage 0. The outputs from stage i are used as the inputs to stage i+1.

The NAND gate is logically complete so any arbitrary boolean function can be implemented by using an appropriately connected collection oi them. The clock signals are now raised in turn. The output of "stage 0" is assumed to be valid prior to the start of operations, and clock 0 is unspecified (for now). Clock 1 is raised, and the outputs from stage 1 become valid. Clock 2 is raised, and the outputs from stage 2 become valid. This sequence continues, until clock N is raised, and the outputs from stage N become valid.

A single stage in the N level circuit could in principle implement an arbitrary boolean switching function by an appropriate arrangement of switches. This might require an exponential number of switches and so multiple stages appear to be useful.

Also, although for simplicity we have assumed that the outputs of stage i are used directly only in stage i+1, they could equally be used in stages i+2, i+3, etc. We will not consider these additional complexities here because they are secondary to the fundamental issues.

LATCHES

What has been described so far will allow the construction of an N level reversible combinational logic circuit. To be useful, we must also specify a latch which can store the output of the circuit. This can be done by holding charge on a capacitor. The outputs of stage N are connected through switches to a latch stage, stage N+1.

The latch requires its own clock. Such a latch is shown in . Initially, the latch clock is 0 and allows charge to move freely into the capacitor. When the output from stage N is valid, the latch clock goes high (changes to logic 1) and prevents charge from escaping. Clock N can now go low (to logic 0) which will make the output from stage N also go low, but the output of the latch will not be effected by changes in the output from stage N because it is cut off.

Of course, having once "locked up" the outputs in the latches, we cannot release the information in the latches. If we should again set the clock latch low, allowing charge to flow off the capacitor, then we would be turning on a switch when there was a voltage across it. This is forbidden, and so the problem of removing information from a latch is non-trivial.

UNWRITING

Rather than simply "erasing" the information on the latch by letting the latch clock go low, we need to "unwrite" the information. Unwriting differs fundamentally from erasing. When we erase, we do not know the state of the latch, we simply turn on the switch holding the charge on the latch and let the charge flow away.

Whether the latch held a "0" or a "1" is immaterial. In the process, we must necessarily dissipate energy because erasing information is irreversible. By contrast, when we "unwrite" a latch we actually have an input signal that tells us the state of the latch. We can set the latch clock low and connect the input signal and the value stored on the latch.

Because (by definition) the input signal is in the same logical state as the latch, there is no voltage across the switch when we turn it on. We can now let the input signal go to 0 slowly, thus draining charge from the latch.We can indefinitely reduce the energy dissipated by this process by slowing the speed with which we drain charge from the latch.

Summary of Combinational Logic

In our earlier description, we left the source of valid input a mystery. The outputs of stage 0 held valid data, and how this happy state of affairs came about was unspecified. Now we can rectify this omission: stage 0 is a latch. This implies that clock 0 must be kept high to prevent the charges in stage 0 from leaking away.

Further, stage N+1 is a latch, and holds the output. We still leave the mechanism that initially put proper charge levels into the capacitors of stage 0 a mystery, but it is now clear that stage 0 will continue to hold a valid signal until clock 0 is lowered and the charge is allowed to escape.

The latches of stage 0 will also be called the input latches or simply the input, while the latches of stage N+1 will also be called the output latches or simply the output.

We can now specify our initial state as one in which appropriate charges are held on the input, clock 0 is high (to prevent charge from the input latches from leaking away), while clocks 1 through N are held low; as is clock signal N+1 (that is, there are no charges held on the output latches and the switch connecting the output latches to the final logic stage is turned on).

We raise clock signals 1 through N in turn, computing the values of each logic level in turn. Finally, we raise clock signal N+1, thus trapping the output signals in the output latches. We are then free to lower clock signal N, then N–1, then N–2, etc. until we have finally worked our way back to the first logic level.

When the computation is done, all clock signals from 1 through N are held low, while clock signals 0 and N+1 remain high. Charge is trapped in stage 0 corresponding to the input, and charge is trapped in stage N+1 corresponding to the output. The outputs of stage 0 and N+1 are valid. All other outputs are 0.ronic Logic.

Chapter 7

Applications of Nanotechnology

With nanotechnology, a large set of materials and improved products rely on a change in the physical properties when the feature sizes are shrunk. Nanoparticles for example take advantage of their dramatically increased surface area to volume ratio. Their optical properties, e. g. fluorescence, become a function of the particle diameter. When brought into a bulk material, nanoparticles can strongly influence the mechanical properties of the material, like stiffness or elasticity.

For example, traditional polymers can be reinforced by nanoparticles resulting in novel materials which can be used as lightweight replacements for metals. Therefore, an increasing societal benefit of such nanoparticles can be expected. Such nanotechnologically enhanced materials will enable a weight reduction accompanied by an increase in stability and an improved functionality. There are many applications of nano technology, few of them are shown here.

MEDICINE

The biological and medical research communities have exploited the unique properties of nanomaterials for various applications (e. g., contrast agents for cell imaging and therapeutics for treating cancer). Terms such as *biomedical nanotechnology, bionanotechnology,* and *nanomedicine* are used to describe this hybrid field. Functionalities can be added to nanomaterials by interfacing them with biological molecules or structures.

The size of nanomaterials is similar to that of most biological molecules and structures; therefore, nanomaterials can be useful for both in vivo and in vitro biomedical research and applications. Thus far, the integration of nanomaterials with biology has led to the development of diagnostic devices, contrast agents, analytical tools, physical therapy applications, and drug delivery vehicles.

Diagnostics

Nanotechnology-on-a-chip is one more dimension of lab-on-a-chip technology. Biological tests measuring the presence or activity of selected substances become quicker, more sensitive and more flexible when certain nanoscale particles are put to work as tags or labels. Magnetic nanoparticles, bound to a suitable antibody, are used to label specific molecules, structures or microorganisms.

Gold nanoparticles tagged with short segments of DNA can be used for detection of genetic sequence in a sample. Multicolor optical coding for biological assays has been achieved by embedding different-sized quantum dots into polymeric microbeads. Nanopore technology for analysis of nucleic acids converts strings of nucleotides directly into electronic signatures.

Drug Delivery

The overall drug consumption and side-effects can be lowered significantly by depositing the active agent in the morbid region only and in no higher dose than needed. This highly selective approach reduces costs and human suffering. An example can be found in dendrimers and nanoporous materials. They could hold small drug molecules transporting them to the desired location. Another vision is based on small electromechanical systems; NEMS are being investigated for the active release of drugs.

Some potentially important applications include cancer treatment with iron nanoparticles or gold shells. A targeted or personalized medicine reduces the drug consumption and treatment expenses resulting in an overall societal benefit by reducing the costs to the public health system. Nanotechnology is also opening up new opportunities in implantable delivery systems, which are often preferable to the use of injectable drugs, because the latter frequently display first-order kinetics (the blood concentration goes up rapidly, but drops exponentially over time). This rapid rise may cause difficulties with toxicity, and drug efficacy can diminish as the drug concentration falls below the targeted range.

Tissue Engineering

Nanotechnology can help to reproduce or to repair damaged tissue. This so called "tissue engineering" makes use of artificially stimulated cell proliferation by using suitable nanomaterial-based scaffolds and growth factors. Tissue engineering might replace today's conventional treatments like organ transplants or artificial implants. On the other hand, tissue engineering is closely related to the ethical debate on human stem cells and its ethical implications.

CHEMISTRY AND ENVIRONMENT

Chemical catalysis and filtration techniques are two prominent examples where nanotechnology already plays a role. The synthesis provides novel materials with tailored features and chemical properties: for example, nanoparticles with a distinct chemical surrounding (ligands), or specific optical properties. In this sense, chemistry is indeed a basic nanoscience.

In a short-term perspective, chemistry will provide novel "nanomaterials" and in the long run, superior processes such as "self-assembly" will enable energy and time preserving strategies. In a sense, all chemical synthesis can be understood in terms of nanotechnology, because of its ability to manufacture certain molecules. Thus, chemistry forms a base for nanotechnology providing tailor-made molecules, polymers, etcetera, as well as clusters and nanoparticles.

Catalysis

Chemical catalysis benefits especially from nanoparticles,

due to the extremely large surface to volume ratio. The application potential of nanoparticles in catalysis ranges from fuel cell to catalytic converters and photocatalytic devices. Catalysis is also important for the production of chemicals.

Platinum nanoparticles are now being considered in the next generation of automotive catalytic converters because the very high surface area of nanoparticles could reduce the amount of platinum required.

However, some concerns have been raised due to experiments demonstrating that they will spontaneously combust if methane is mixed with the ambient air. Ongoing research at the Centre National de la Recherche Scientifique (CNRS) in France may resolve their true usefulness for catalytic applications. Nanofiltration may come to be an important application, although future research must be careful to investigate possible toxicity.

Filtration

A strong influence of nanochemistry on waste-water treatment, air purification and energy storage devices is to be expected. Mechanical or chemical methods can be used for effective filtration techniques.

One class of filtration techniques is based on the use of membranes with suitable hole sizes, whereby the liquid is pressed through the membrane. Nanoporous membranes are suitable for a mechanical filtration with extremely small pores smaller than 10 nm ("nanofiltration") and may be composed of nanotubes. Nanofiltration is mainly used for the removal of ions or the separation of different fluids.

On a larger scale, the membrane filtration technique is named ultrafiltration, which works down to between 10 and 100 nm.

One important field of application for ultrafiltration is medical purposes as can be found in renal dialysis. Magnetic nanoparticles offer an effective and reliable method to remove heavy metal contaminants from waste water by making use of magnetic separation techniques. Using nanoscale particles increases the efficiency to absorb the contaminants and is comparatively inexpensive compared to traditional precipitation and filtration methods.

Some water-treatment devices incorporating nanotechnology are already on the market, with more in development. Low-cost nanostructured separation membranes methods have been shown to be effective in producing potable water in a recent study.

ENERGY

The most advanced nanotechnology projects related to energy are: storage, conversion, manufacturing improvements by reducing materials and process rates, energy saving (by better thermal insulation for example), and enhanced renewable energy sources.

Reduction of Energy Consumption

A reduction of energy consumption can be reached by better insulation systems, by the use of more efficient lighting or combustion systems, and by use of lighter and stronger materials in the transportation sector. Currently used light bulbs only convert approximately 5per cent of the electrical energy into light. Nanotechnological approaches like light-emitting diodes (LEDs) or quantum caged atoms (QCAs) could lead to a strong reduction of energy consumption for illumination.

Increasing the Efficiency of Energy Production

Today's best solar cells have layers of several different semiconductors stacked together to absorb light at different energies but they still only manage to use 40 per cent of the Sun's energy.

Commercially available solar cells have much lower efficiencies (15-20per cent). Nanotechnology could help increase the efficiency of light conversion by using nanostructures with a continuum of bandgaps. The degree of efficiency of the internal combustion engine is about 30-40per cent at the moment.

Nanotechnology could improve combustion by designing

specific catalysts with maximized surface area. In 2005, scientists at the University of Toronto developed a spray-on nanoparticle substance that, when applied to a surface, instantly transforms it into a solar collector.

The use of more Environmentally Friendly Energy Systems

An example for an environmentally friendly form of energy is the use of fuel cells powered by hydrogen, which is ideally produced by renewable energies.

Probably the most prominent nanostructured material in fuel cells is the catalyst consisting of carbon supported noble metal particles with diameters of 1-5 nm. Suitable materials for hydrogen storage contain a large number of small nanosized pores. Therefore many nanostructured materials like nanotubes, zeolites or alanates are under investigation. Nanotechnology can contribute to the further reduction of combustion engine pollutants by nanoporous filters, which can clean the exhaust mechanically, by catalytic converters based on nanoscale noble metal particles or by catalytic coatings on cylinder walls and catalytic nanoparticles as additive for fuels.

Recycling of Batteries

Because of the relatively low energy density of batteries the operating time is limited and a replacement or recharging is needed. The huge number of spent batteries and accumulators represent a disposal problem. The use of batteries with higher energy content or the use of rechargeable batteries or supercapacitors with higher rate of recharging using nanomaterials could be helpful for the battery disposal problem.

INFORMATION AND COMMUNICATION

Current high-technology production processes are based on traditional top down strategies, where nanotechnology has already been introduced silently. The critical length scale of integrated circuits is already at the nanoscale (50 nm and below) regarding the gate length of transistors in CPUs or DRAM devices.

Memory Storage

Electronic memory designs in the past have largely relied on the formation of transistors however research into crossbar switch based electronics have offered an alternative using reconfigurable interconnections between vertical and horizontal wiring arrays to create ultra high density memories. Two leaders in this area are Nantero which has developed a carbon nanotube based crossbar memory called Nano-RAM and Hewlett-Packard which has proposed the use of memristor material as a future replacement of Flash memory.

Novel Semiconductor Devices

An example of such novel devices is based on spintronics. The dependence of the resistance of a material (due to the spin of the electrons) on an external field is called magnetoresistance. This effect can be significantly amplified (GMR - Giant Magneto-Resistance) for nanosized objects, for example when two ferromagnetic layers are separated by a nonmagnetic layer, which is several nanometers thick (e. g. Co-Cu-Co).

The GMR effect has led to a strong increase in the data storage density of hard disks and made the gigabyte range possible. The so called tunneling magnetoresistance (TMR) is very similar to GMR and based on the spin dependent tunneling of electrons through adjacent ferromagnetic layers. Both GMR and TMR effects can be used to create a non-volatile main memory for computers, such as the so called magnetic random access memory or MRAM.

In 1999, the ultimate CMOS transistor developed at the Laboratory for Electronics and Information Technology in Grenoble, France, tested the limits of the principles of the MOSFET transistor with a diameter of 18 nm (approximately 70 atoms placed side by side). This was almost one tenth the size of the smallest industrial transistor in 2003 (130 nm in 2003, 90 nm in 2004, 65 nm in 2005 and 45 nm in 2007).

It enabled the theoretical integration of seven billion junctions on a €1 coin. However, the CMOS transistor, which was created in 1999, was not a simple research experiment to study how CMOS technology functions, but rather a demonstration of how this technology functions now that we ourselves are getting ever closer to working on a molecular scale. Today it would be impossible to master the coordinated assembly of a large number of these transistors on a circuit and it would also be impossible to create this on an industrial level.

Novel Optoelectronic Devices

In the modern communication technology traditional analog electrical devices are increasingly replaced by optical or optoelectronic devices due to their enormous bandwidth and capacity, respectively. Two promising examples are photonic crystals and quantum dots. Photonic crystals are materials with a periodic variation in the refractive index with a lattice constant that is half the wavelength of the light used.

They offer a selectable band gap for the propagation of a certain wavelength, thus they resemble a semiconductor, but for light or photons instead of electrons. Quantum dots are nanoscaled objects, which can be used, among many other things, for the construction of lasers. The advantage of a quantum dot laser over the traditional semiconductor laser is that their emitted wavelength depends on the diameter of the dot. Quantum dot lasers are cheaper and offer a higher beam quality than conventional laser diodes.

Displays

The production of displays with low energy consumption could be accomplished using carbon nanotubes (CNT). Carbon nanotubes be electrically conductive and due to their small diameter of several nanometers, they can be used as field emitters with extremely high efficiency for field emission displays (FED). The principle of operation resembles that of the cathode ray tube, but on a much smaller length scale.

Quantum Computers

Entirely new approaches for computing exploit the laws of quantum mechanics for novel quantum computers, which enable the use of fast quantum algorithms. The Quantum computer will have quantum bit memory space termed qubit for several computations at the same time.

HEAVY INDUSTRY

An inevitable use of nanotechnology will be in heavy industry.

Aerospace

Lighter and stronger materials will be of immense use to aircraft manufacturers, leading to increased performance. Spacecraft will also benefit, where weight is a major factor. Nanotechnology thereby helps to reduce the size of equipments used.

Hang gliders halve their weight while increasing their strength and toughness through the use of nanotech materials. Nanotech is lowering the mass of supercapacitors that will increasingly be used to give power to assistive electrical motors for launching hang gliders off flatland to thermal-chasing altitudes.

Refineries

Using nanotech applications, refineries producing materials such as steel and aluminium will be able to remove any impurities in the materials they create.

Vehicle Manufacturers

Much like aerospace, lighter and stronger materials will be useful for creating vehicles that are both faster and safer. Combustion engines will also benefit from parts that are more hard-wearing and more heat-resistant.

CONSUMER GOODS

Nanotechnology is already impacting the field of consumer goods, providing products with novel functions ranging from easy-to-clean to scratch-resistant. Modern textiles are wrinkle-resistant and stain-repellent; in the mid-term clothes will become "smart", through embedded "wearable electronics". Already in use are different nanoparticle improved products. Especially in the field of cosmetics, such novel products have a promising potential.

Foods

Nanotechnology can be applied in the production, processing, safety and packaging of food. A nanocomposite coating process could improve food packaging by placing anti-microbial agents directly on the surface of the coated film. Nanocomposites could increase or decrease gas permeability of different fillers as is needed for different products. They can also improve the mechanical and heat-resistance properties and lower the oxygen transmission rate. Research is being performed to apply nanotechnology to the detection of chemical and biological substances for sensing biochemical changes in foods.

Household

The most prominent application of nanotechnology in the household is self-cleaning or "easy-to-clean" surfaces on ceramics or glasses. Nanoceramic particles have improved the smoothness and heat resistance of common household equipment such as the flat iron.

Optics

The first sunglasses using protective and antireflective ultrathin polymer coatings are on the market. For optics, nanotechnology also offers scratch resistant surface coatings based on nanocomposites. Nano-optics could allow for an increase in precision of pupil repair and other types of laser eye surgery.

Textiles

The use of engineered nanofibers already makes clothes water- and stain-repellent or wrinkle-free. Textiles with a nanotechnological finish can be washed less frequently and at lower temperatures. Nanotechnology has been used to integrate tiny carbon particles membrane and guarantee full-surface protection from electrostatic charges for the wearer.

Many other applications have been developed by research institutions such as the Textiles Nanotechnology Laboratory at Cornell University

Cosmetics

One field of application is in sunscreens. The traditional chemical UV protection approach suffers from its poor long-term stability. A sunscreen based on mineral nanoparticles such as titanium dioxide offer several advantages. Titanium oxide nanoparticles have a comparable UV protection property as the bulk material, but lose the cosmetically undesirable whitening as the particle size is decreased.

ENERGY APPLICATIONS OF NANOTECHNOLOGY

Over the past few decades, the fields of science and engineering have been seeking to develop new and improved types of energy technologies that have the capability of improving life all over the world. In order to make the next leap forward from the current generation of technology, scientists and engineers have been developing Energy Applications of Nanotechnology. Nanotechnology, a new field in science, is any technology that contains components smaller than 100 nanometers. For scale, a single virus particle is about 100 nanometers in width.

An important subfield of nanotechnology related to energy is nanofabrication. Nanofabrication is the process of designing and creating devices on the nanoscale. Creating devices smaller than 100 nanometers opens many doors for the development of new ways to capture, store, and transfer energy. The inherent level of control that nanofabrication could give scientists and engineers would be critical in providing the capability of solving many of the problems that the world is facing today related to the current generation of energy technologies.

People in the fields of science and engineering have already begun developing ways of utilizing nanotechnology for the development of consumer products. Benefits already observed from the design of these products are an increased efficiency of lighting and heating, increased electrical storage capacity, and a decrease in the amount of pollution from the use of energy. Benefits such as these make the investment of capital in the research and development of nanotechnology a top priority.

CONSUMER PRODUCTS

Recently, previously established and entirely new companies such as BetaBatt, Inc. and Oxane Materials are focusing on nanomaterials as a way to develop and improve upon older methods for the capture, transfer, and storage of energy for the development of consumer products.

Conserve, a product developed by the Dais Analytic Corporation, uses nanoscale polymer membranes to increase the efficiency of heating and cooling systems and has already proven to be a lucrative design. The polymer membrane was specifically configured for this application by selectively engineering the size of the pores in the membrane to prevent air from passing, while allowing moisture to pass through the membrane. Polymer membranes can be designed to selectively allow particles of one size and shape to pass through while preventing others of different dimensions. This makes for a powerful tool that can be used in consumer products from biological weapons protection to industrial chemical separations.

A New York based company called Applied NanoWorks, Inc. has been developing a consumer product that utilizes LED technology to generate light. Light-emitting diodes or LEDs, use only about 10per cent of the energy that a typical incandescent or fluorescent light bulb use and typically lasts much longer, which makes them a viable alternative to traditional light bulbs. While LEDs have been around for decades, this company and others like it have been developing a special variant of LED called the white LED.

White LEDs consist of semi-conducting organic layers that are only about 100 nanometers in distance from each other and are placed between two electrodes, which create an anode, and a cathode. When voltage is applied to the system, light is generated when electricity passes through the two organic layers. This is called electroluminescence. The semiconductor properties of the organic layers are what allow for the minimal amount of energy necessary to generate light. In traditional light bulbs, a metal filament is used to generate light when electricity is run through the filament. Using metal generates a great deal of heat and therefore lowers efficiency.

Research for longer lasting batteries has been an ongoing process for years. Researchers have now begun to utilize nanotechnology for battery technology. mPhase Technologies in conglomeration with Rutgers University and Bell Laboratories have utilized nanomaterials to alter the wetting behaviour of the surface where the liquid in the battery lies to spread the liquid droplets over a greater area on the surface and therefore have greater control over the movement of the droplets. This gives more control to the designer of the battery. This control prevents reactions in the battery by separating the electrolytic liquid from the anode and the cathode when the battery is not in use and joining them when the battery is in need of use.

ECONOMIC BENEFITS

The relatively recent shift toward using nanotechnology with respect to the capture, transfer, and storage of energy has and will continue to have many positive economic impacts on society. The control of materials that nanotechnology offers to scientists and engineers of consumer products is one of the most important aspects of nanotechnology. This allows for an improved efficiency of products across the board.

A major issue with current energy generation is the loss of efficiency from the generation of heat as a by-product of the process. A common example of this is the heat generated by the internal combustion engine.

The internal combustion engine loses about 64per cent of the energy from gasoline as heat and an improvement of this alone could have a significant economic impact. However, improving the internal combustion engine in this respect has proven to be extremely difficult without sacrificing performance. Improving the efficiency of fuel cells through the use of nanotechnology appears to be more plausible by using molecularly tailored catalysts, polymer membranes, and improved fuel storage.

In order for a fuel cell to operate, particularly of the hydrogen variant, a noble-metal catalyst (usually platinum, which is very expensive) is needed to separate the electrons from the protons of the hydrogen atoms. However, catalysts of this type are extremely sensitive to carbon monoxide reactions. In order to combat this, alcohols or hydrocarbons compounds are used to lower the carbon monoxide concentration in the system. This adds an additional cost to the device. Using nanotechnology, catalysts can be designed through nanofabrication that are much more resistant to carbon monoxide reactions, which improves the efficiency of the process and may be designed with cheaper materials to additionally lower costs.

Fuel cells that are currently designed for transportation need rapid start-up periods for the practicality of consumer use. This process puts a lot of strain on the traditional polymer electrolyte membranes, which decreases the life of the membrane requiring frequent replacement. Using nanotechnology, engineers have the ability to create a much more durable polymer membrane, which addresses this problem. Nanoscale polymer membranes are also much more efficient in ionic conductivity. This improves the efficiency of the system and decreases the time between replacements, which lowers costs.

Another problem with contemporary fuel cells is the storage of the fuel. In the case of hydrogen fuel cells, storing the hydrogen in gaseous rather than liquid form improves the efficiency by 5per cent. However, the materials that we currently have available to us significantly limit fuel storage due to low stress tolerance and costs. Scientists have come up with an answer to this by using a nanoporous styrene material (which is a relatively inexpensive material) that when super-cooled to around -196°C, naturally holds on to hydrogen atoms and when heated again releases the hydrogen for use.

CAPACITORS: THEN AND NOW

For decades, scientists and engineers have been attempting to make computers smaller and more efficient. A crucial component of computers are capacitors. A capacitor is a device that is made of a pair of electrodes separated by an insulator that each stores an opposite charge. A capacitor stores a charge when it is removed from the circuit that it is connected to; the charge is released when it is replaced back into the circuit. Capacitors have an advantage over batteries in that they release their charge much more quickly than a battery.

Traditional or foil capacitors are composed of thin metal conducting plates separated by an electrical insulator, which are then stacked or rolled and placed in a casing. The problem with a traditional capacitor such as this is that they limit how small an engineer can design a computer. Scientists and engineers have since turned to nanotechnology for a solution to the problem.

Using nanotechnology, researchers developed what they call "ultracapacitors. "An ultracapacitor is a general term that describes a capacitor that contains nanocomponents. Ultracapacitors are being researched heavily because of their high density interior, compact size, reliability, and high capacitance. This decrease in size makes it increasingly possible to develop much smaller circuits and computers.

Ultracapacitors also have the capability to supplement batteries in hybrid vehicles by providing a large amount of energy during peak acceleration and allowing the battery to supply energy over longer periods of time, such as during a constant driving speed. This could decrease the size and weight of the large batteries needed in hybrid vehicles as well as take additional stress off the battery. However, as of now, the combination of ultracapacitors and a battery is not cost effective due to the need of additional DC/DC electronics to coordinate the two.

Nanoporous carbon aerogel is one type of material that is being utilized for the design of ultracapacitors. These aerogels have a very large interior surface area and can have its properties altered by changing the pore diameter and distribution along with adding nanosized alkali metals to alter its conductivity.

Carbon nanotubes are another possible material for use in an ultracapacitor. Carbon nanotubes are created by vaporizing carbon and allowing it to condense on a surface. When the carbon condenses, it forms a nanosized tube composed of carbon atoms. This tube has a high surface area, which increases the amount of charge that can be stored. The low reliability and high cost of using carbon nanotubes for ultracapacitors is currently an issue of research.

In a study concerning ultracapacitors or supercapacitors, researchers at the Sungkyunkwan University in the Republic of Korea explored the possibility of increasing the capacitance of electrodes through the addition of fluorine atoms to the walls of carbon nanotubes. Carbon nanotubes are an increasing form of capacitors due to their superb chemical stability, high conductivity, light mass, and their large surface area. These researchers fluorinated single-walled carbon nanotubes (SWCNTs) at high temperatures to bind fluorine atoms to the walls. The attached fluorine atoms changed the non-polar nanotubes to become polar molecules.

This can be attributed to the charge transfer from the fluorine. This created dipole-dipole layers along the carbon nanotube walls. Testing of these fluorinated SWCNTs against normal state SWCNTs showed a difference in capacitance. It was determined that the fluorinated SWCNTs are advantageous in fabricating electrodes for capacitors and improve the wettability with aqueous electrolytes, which promotes the overall performance of supercapacitors.

While this study brought to knowledge a more efficient example of capacitors, little is known about this new supercapacitor, large scale synthesis is lacking and is necessary for any massive production, and preparation conditions are quite tedious in achieving the final product.

THEORY OF CAPACITANCE

Understanding the concept of capacitance can be helpful in understanding why nanotechnology is such a powerful tool for the design of higher energy storing capacitors. A capacitor's capacitance (C) or amount of energy stored is equal to the amount of charge (Q) stored on each plate divided by the voltage (V) between the plates.

Another representation of capacitance is that capacitance (C) is approximately equal to the permittivity (å) of the dielectric times the area (A) of the plates divided by the distance (d) between them. Therefore, capacitance is proportional to the surface area of the conducting plate and inversely proportional to the distance between the plates.

Using carbon nanotubes as an example, a property of carbon nanotubes is that they have a very high surface area to store a charge. Using the above proportionality that capacitance (C) is proportional to the surface area (A) of the conducting plate; it becomes obvious that using nanoscaled materials with high surface area would be great for increasing capacitance. The other proportionality described above is that capacitance (C) is inversely proportional to the distance (d) between the plates. Using nanoscaled plates such as carbon nanotubes with nanofabrication techniques, gives the capability of decreasing the space between plates which again increases capacitance.

HOW NANOTECHNOLOGY IMPACTS OUR LIVES

NANOCOMPOSITES

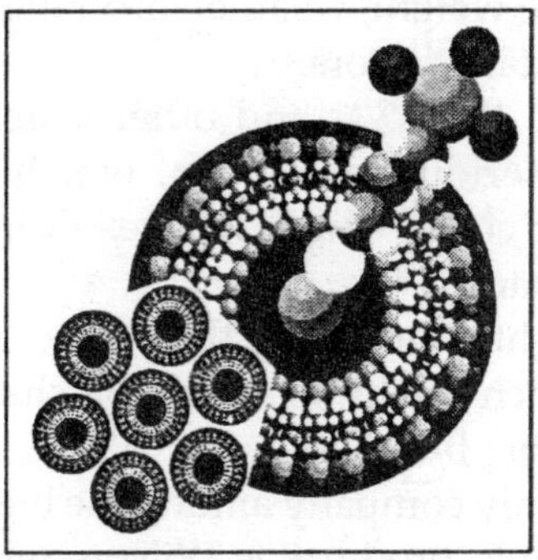

Fig. Self-Assembled Monolayers on Mesoporous Supports

Researchers at Pacific Northwest National Laboratory have developed a coating process to make sponge-like silica latch onto toxic metals in water.

Self-Assembled Monolayers on Mesoporous Supports easily captures such metals as lead and mercury, which are then recovered for reuse or contained in-place forever. PNNL One example of a SAMMS nanocomposite (Self-Assembled Monolayers on Mesoporous Supports).

An hexagonally close-packed cluster of tubular pores (end view) is shown in the foreground. A single pore, in this case coated with a mercaptopropylsiloxy monolayer. A model of one surfactant molcule is also shown.

A plastic nanocomposite is being used for "step assists" in the GM Safari and Astro Vans. It is scratch-resistant, light-weight, and rust-proof, and generates improvements in strength and reductions in weight, which lead to fuel savings and increased longevity.

And in 2001, Toyota started using nanocomposites in a bumper that makes it 60 per cent lighter and twice as resistant to denting and scratching.

Impact: Will likely be used on other GM and Toyota models soon, and in other areas of their vehicles, as well as the other auto manufactures, lowering weight, increasing milage, and creating longer-lasting autos.

Likely to impact repair shops (fewer repairs needed) and auto insurance companies (fewer claims). Will also likely soon be seen everywhere weight, weather-proofing, durability, and strength are important factors.

Expect NASA, the ESA, and other space-faring organizations to take a serious look, soon, which will eventually result in lower lift costs, which will result in more material being lifted into space.

Strong and light sounds like the perfect recipe for a golf club, of course, which is why so many golf club manufacturers are now devoting big bucks to nano R&D.

Thus far, the only company among the big boys to convert research into tangible products is Wilson, which offers three drivers, a fairway wood, four balls and even a golf bag made using nano-materials. The Chicago-based firm began looking into the potential of nanotechnology six years ago, but didn't come out with a product - the NCODE series of tennis rackets - until 2004.

Later in the year, it started shipping drivers and fairway woods, the Pd5, Dd5, Td5 (MSRP - $300) and FwC (MSRP - $200), whose crowns are constructed with nano carbon which, Wilson claims, creates a low-density, high-strength clubhead. Angus Moir, global business director, says use of nano materials sets Wilson apart from the competition. "They make our products more user friendly," he says.

NANOCRYSTALS

Examples: "Metal nanocrystals might be incorporated into car bumpers, making the parts stronger, or into aluminum, making it more wear resistant. Metal nanocrystals might be used to produce bearings that last longer than their conventional counterparts, new types of sensors and components for computers and electronic hardware.

Nanocrystals of various metals have been shown to be 100 per cent, 200 per cent and even as much as 300 per cent harder than the same materials in bulk form. Because wear resistance often is dictated by the hardness of a metal, parts made from nanocrystals might last significantly longer than conventional parts. " Six different quantum dot solutions are shown, excited with a long-wave UV lamp.

Quantum dots are molecular-scale optical beacons. Qdot nanocrystals behave like molecular LEDs (light emitting diodes) by "lighting up" biological binding events with a broad palette of applied colors. Smith & Nephew markets an antimicrobial dressing covered with nanocrystalline silver (A patented Technology of NUCRYST Pharmaceuticals). The nanocrystalline coating of silver rapidly kills a broad spectrum of bacteria in as little as 30 minutes.

Nanocrystals are an ideal light harvester in photovoltaic devices. They absorb sunlight more strongly than dye molecules or bulk semiconductor material, therefore high optical densities can be achieved while maintaining the requirement of thin films. Perfectly crystalline CdSe nanocrystals are also an artificial reaction centre, separating the electron hole pair on a femtosecond timescale.

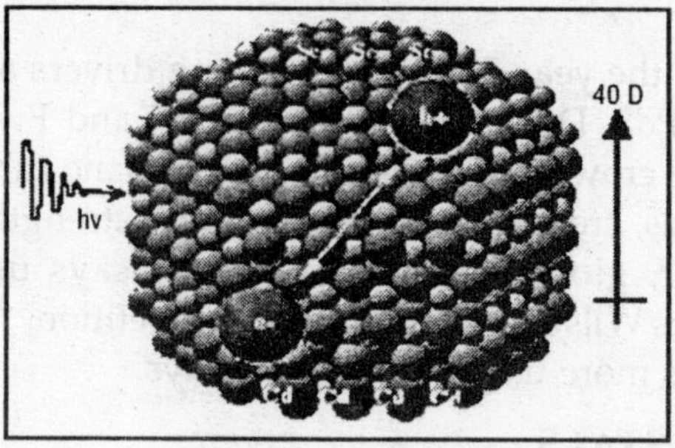

Fig. Nanocrystals

Fluorescent nanocrystals have several advantages over organic dye molecules as fluorescent markers in biology. They are incredibly bright and do not photodegrade. Drug-conjugated nanocrystals attach to the protein in an extracellular fashion, enabling movies of protein trafficking. They also form the basis of a high-throughput fluorescence assay for drug discovery. "

NANOPARTICLES

Stain-repellent Eddie Bauer Nano-Care khakis, with surface fibers of 10 to 100 nanometers, uses a process that coats each fibre of fabric with "nano-whiskers. "Developed by Nano-Tex, a Burlington Industries subsidiary. Dockers also makes khakis, a dress shirt and even a tie treated with what they call "Stain Defender", another example of the same nanoscale cloth treatment.

Dry cleaners, detergent and stain-removal makers, carpet and furniture makers, window covering makers.

BASF's annual sales of aqueous polymer dispersion products amount to around $1. 65 billion. All of them contain polymer particles ranging from ten to several hundred nanometers in size. Polymer dispersions are found in exterior paints, coatings and adhesives, or are used in the finishing of paper, textiles and leather. Nanotechnology also has applications in the food sector. Many vitamins and their precursors, such as carotinoids, are insoluble in water. However, when skillfully produced and formulated as nanoparticles, these substances can easily be mixed with cold water, and their bioavailability in the human body also increases.

Many lemonades and fruit juices contain these specially formulated additives, which often also provide an attractive colour. In the cosmetics sector, BASF has for several years been among the leading suppliers of UV absorbers based on nanoparticulate zinc oxide. Incorporated in sun creams, the small particles filter the high-energy radiation out of sunlight. Because of their tiny size, they remain invisible to the naked eye and so the cream is transparent on the skin.

Sunscreens are utilizing nanoparticles that are extremely effective at absorbing light, especially in the ultra-violet (UV) range. Due to the particle size, they spread more easily, cover better, and save money since we use less. And they are transparent, unlike traditional screens which are white. These sunscreens are so successful that by 2001 they had captured 60per cent of the Australian sunscreen market.

Makers of sunscreen have to convert to using nano-particles. And other product manufactures, like packaging makers, will find ways to incorporate them into packages to reduces UV exposure and subsequent spoilage. The $480B packaging and $300B plastics industries will be directly effected. Using aluminum nanoparticles, Argonide has created rocket propellants that burn at double the rate. They also produce copper nanoparticles that are incorporated into automotive lubricant to reduce engine wear.

AngstroMedica has produced a nanoparticulate-based synthetic bone. "Human bone is made of a calcium and phosphate composite called Hydroxyapatite.

By manipulation calcium and phosphate at the molecular level, we have created a patented material that is identical in structure and composition to natural bone. This novel synthetic bone can be used in areas where natural bone is damaged or removed, such as in the in the treatment of fractures and soft tissue injuries. "

Index